A MANAGER'S GUIDE TO QUALITY AND RELIABILITY

A Manager's Guide to Quality and Reliability

RUPERT GEDYE
Lecturer, Management Studies
Bath University of Technology
England

1968
JOHN WILEY & SONS
London — New York — Sydney

Copyright 1968 by John Wiley & Sons Ltd.
ALL RIGHTS RESERVED

No part of this book may be
reproduced by any means, nor
transmitted, nor translated
into a machine language without
the written permission of the publisher

Library of Congress Catalog card number 68–21640

SBN 471 29500 0

Printed in Great Britain by
Unwin Brothers Limited
Woking and London

Foreword

AT A CRITICAL stage in the play 'Peter Pan', Tinkerbell the fairy rather stupidly drinks poison. All seems lost, no remedy seems appropriate to save her and science is helpless. At this critical moment Peter invites all those in the audience who believe in fairies to clap their hands. Everyone does so and, without recourse to any science at all, Tinkerbell is saved.

As year succeeds year, our national mood seems to become more psychotic. We beat our breasts and bemoan our failures and all seems lost. At the moment of greatest darkness an exhortatory cry comes from the Government, and all those who believe in productivity, or quality and reliability, are invited to clap their hands throughout a whole productivity year or a whole quality and reliability year. One feature of these annual events is that they seem to have rather more emotion than rationality associated with them. There is no definition of what is meant by the terms used, nor is there any quantitative evaluation of the effect on the G.N.P. or balance of payments of the various degrees of success of these years.

Perhaps the greatest value of these national events is that they draw the attention of the backwood areas of management to the availability of modern techniques and bring out comparisons between firms in the same industry in our own country, and the general picture in this country compared with our foreign rivals. If this is done, then these national events are justified since they may well stir the consciences, or even the minds, of management.

This book is not exhortatory in tone but is rather a serious contribution to the methodology of quality and reliability and seeks to place this problem within the general context of industrial activity. It is written by a man who has many years' experience in the top levels of industry, who has had personal responsibility for designing and implementing such schemes, and who is now sharing his long experience with students, as a member of staff of one of our universities.

The Manager's Guide series, of which this book forms a part, is an attempt to provide managers with a series of brief, non-technical introductions to the different specialisms which seek to improve the processes of decision-making and control in industry. It is hoped that each of these books will enable the executive to know whether this is a topic about which he should worry and, if so, how he should best develop his interest in the topic. They are not simply technical guides although, of course, it is, in general, difficult to understand the relevance of a technical subject without taking the trouble to understand a little about the techniques involved.

It is in this spirit that I commend this present book to those who would like to know, in a rational way, what quality and reliability, as a technical subject, has to offer and what is involved in an organization developing competence in this field.

B. H. P. RIVETT

Sussex

January, 1968

Preface

THE RAPID development of the techniques of quality and reliability control in recent years has been an important feature of modern industry.

A reputation for sound and reliable goods produced at an economic cost is vital to any firm which is to grow and prosper. Success in this aim depends on the extent to which it pervades all aspects of the business. The quality specialist can provide a service, but the results achieved depend on the manager and those he controls, down to the primary producer. This book is written primarily for the manager, to guide him as to the decisions involved in introducing or developing improved quality and reliability techniques and to enable him to understand what the specialist can do to help him in his responsibilities. It should also be useful to the management student, not only in such courses as the Certificate and Diploma of the Institute of Works Managers and the Diploma of Management Studies, but as an introduction to what will be an important aspect of his future career.

As in all management techniques, human relations play a vital part and this has been stressed and illustrated throughout the book and particularly in Chapter 3, which includes a discussion on incentives and their relation to quality. The book concludes with a review of the wider economic aspects of quality and reliability. It is hoped that these features will not only be of interest to the industrial management, but to trades union officials and civil servants.

In writing this book I have received help from many sources for which I am very grateful in addition to the acknowledgements given overleaf. First I should like to thank Professor Patrick Rivett, for his suggestion that I should write the book, for including a Foreword, and for the interest and encouragement he has given. I am also indebted to the National Productivity Council and the Production Engineering Research Association for information gained during their National Quality and Reliability Year conferences. I am obliged to a number of my former colleagues at Nairn-Williamson Ltd. for their co-operation, and particularly David Jackson and Marion Garbutt for help with statistical case studies and diagrams and Maureen Bethell for her valuable assistance including typing the manuscript, much of it in her own time.

Bath University of Technology
September, 1967

Acknowledgements

THANKS ARE DUE to the following institutions and individuals for help given and/or permission to quote material or publish case studies:

The Crown Agents

W. D. Farrington, C.B.E.,
Chief Inspecting Engineer

Domco Vinyls Ltd., Brantford, Ontario

James P. Coleman,
Plant Manager

Marks & Spencer Ltd.

H. L. Fiebelmann, B.Sc.(Eng.), A.M.I.Prod.E., M.I.M.C.,
Production Engineer

Nairn-Williamson Ltd.

W. Roxburgh, D.F.C.,
Group Managing Director

Production Engineering Research Association

P. R. Lofthouse,
Conference Organizer

The Rover Company Ltd.

M. T. Witts, C.Eng., M.I.Mech.E.,
Deputy Chief Engineer— Reliability

Spicers (Stationery) Ltd.

D. A. Ross-Stewart,
General Manager

Standard Coated Products Inc., Buchanan, N.Y., U.S.A.

Edgar S. Thompson,
President

Contents

Foreword	v
Preface	vii
1. Introduction	1
2. Inspection and Quality Control	6
3. Some Human Aspects	16
4. Statistical Concepts	24
5. Control Charts	45
6. Design for Quality and Reliability	75
7. Organizing for Quality and Reliability	98
8. Economic Aspects of Quality and Reliability	110
Books Recommended for Further Reading	122
References	124
Index	125

1 Introduction

'The best protection against acceptance of a defective product is, of course, having it made right in the first place.'

E. L. Grant
Statistical Quality Control

A SOUND and reliable product is the foundation of successful manufacturing business. Without it, no amount of hard work, good organization or sales effort can for long produce results. National Quality and Reliability Year, which opened in Britain in October 1966, focused attention on the extreme importance to the national economy of a high reputation for the quality and reliability of its goods and services to the customer, both at home and abroad. Under today's competitive conditions it is essential not only to have a good product, but to manufacture and market it economically. As profitability must be maintained or increased if our businesses are to live and prosper, this implies that costs must be reduced. It is commonly held that quality can only be improved by using better and more costly materials, or by more expensive workmanship. The object of this book is to show how a properly designed and administered quality and reliability system, far from increasing costs and prices, can be a major factor in improved profitability, not only through the expansion of sales, but through a reduction in direct costs. It has been estimated from a wide survey of British industry that quality costs, which we may define as the combined costs of prevention of poor quality, inspection, rectification and rejection, frequently amount to as much as 10% of the company's turnover. When we add to this the costs which can arise from unreliability of goods and services, in the form of increased inventories, the avoidance and corrections of errors, the settling of complaints and the loss of goodwill, it will be seen what a fruitful field for cost saving exists.

What do we understand by quality and reliability? In ordinary speech the word quality means general excellence. To the customer for consumer goods it implies excellence in the properties he expects and looks for, for example appearance, durability and performance. Whether the product possesses these properties depends first of all on whether they were designed into the product, and if so, the extent to which the manufactured product conforms with the designer's specification.

This leads us to the important concepts of quality of design and quality of conformance. The former may be difficult to express in figures, but the latter is usually measurable in terms of the percentage conforming to specification. Reliability is generally understood to mean the probability that a product will give satisfactory service and not break down or fail prematurely under specified or reasonable operating conditions. A well designed and well made article will normally be reliable, so reliability should be regarded as a special facet of quality, rather than something distinct. The term reliability also applies to the degree of dependence customers can place on services and delivery promises.

To be truly effective, a quality and reliability system must begin and end with the customer, who can be regarded either as a potential executioner or as a partner; no effort should be spared to cultivate the latter relationship. It is important to recognize that the customer requires goods consistently up to the standard of quality he expects, coupled with reliability in delivery and at the price he is prepared to pay. It is essential to clear from our minds the idea that cost reduction is only possible at the expense of quality, or that quality improvement must necessarily incur additional cost. First, therefore, we must study the customer's requirements and fully appreciate the conditions under which the product will be used. Next comes good design, i.e. not only design of the right end product, but for the most effective and efficient method of manufacturing it. Correct original design and layout reduces the incidence of manufacturing faults. Having established the design on the basis of a product which is right for the job, free of costs which contribute nothing to saleability and performance, our next task is to ensure that manufacture complies with design, within the limits of tolerance necessary to ensure satisfactory service. If the product is made right, the problems and expense of inspection and rejection will be minimized. In short, good design and method from start to finish should result in a reliable product with the minimum rejections for faulty work or handling damage.

It is essential to appreciate that, however good our methods and layout may be, some variability will occur and it is therefore necessary to control this variability with limits which will not result in the product failing to fulfil its function. Control of variability within acceptable limits may well be far more important to the customer than a general raising of the average quality standard in the absence of such control. A chain is as strong as its weakest link. Thus if we are manufacturing and marketing a product in which strength is an important feature, it will probably be far more important to eliminate the occasional weakness which destroys the product's reputation for reliability than to raise the general standard of strength through use of more costly materials and still retain the high variability which will result in periodical failures. Again, if we are concerned with length of service, a product on which we can be assured of a dependable performance for, say, two to three years may give much greater satisfaction than one designed for 10 years' life which occasionally fails in the first two months. The reputation for reliability and consequent sales of a new car model will be much more affected by the occasional customer who experiences serious trouble or is left stranded on the road within a few months of his purchase than by reports or guarantees that the car is designed to last for a period of years.

The main features of a sound quality and reliability system may conveniently be summed up in the following logical eight-point plan which emerged at the National Conference of Quality and Reliability at Blackpool in November 1966.[1]

(1) Study of customer requirements with due consideration to performance and price.

(2) Satisfactory design of product, or service, thoroughly proved by adequate testing in order to establish its reliability under the conditions to which it will be subjected in use.

(3) Full specification of the requirements of the design

clearly understood by everyone concerned with the constituent parts and of the complete end product or service.

(4) Confirmation that the production, or operating processes, are capable of meeting the design requirements.

(5) Full acceptance, by all those concerned with these processes, of the responsibility for meeting the standards set by the specification.

(6) Check that the product or service conforms with the specification.

(7) Instruction in the use, application and limitations of the product or service.

(8) Study of user experience, feed-back to the department concerned and rapid remedial action.

In this book we shall review many of the techniques whereby a plan of this nature can be implemented and at the same time give substantial savings in direct costs.

Some years ago I was talking to a consultant who was working on an assignment in our floorcovering factory and mentioned quality control. He said, 'What you are doing here is not quality control but physical testing and inspection'. I then asked him what he understood by quality control and he replied, 'Quality control is a mathematical technique which has been very useful in the engineering industry, but would have no application here'. Following this conversation I set out to find all I could about quality control, and soon realized that while his first statement was correct, his second was quite wrong. Variability is a major cause of quality failure, and quality can normally be measured. As statistics is the branch of mathematics which enables us to handle variable measurements, statistical methods are most valuable in controlling quality, but while they were first developed and publicized in engineering, they are equally applicable in a wide range of process industry. As we shall see in later chapters, quality control embraces a wider field and statistical methods, while vital in any good quality control system where measurement or counting is possible, are by no means the whole story.

Our eight-point plan quoted above emphasizes the fact that

the effective control of quality involves all sections of the organization, including market research, design, purchasing, production, inspection and testing, marketing and sales, and all levels from higher management and technical staff to the shop floor. This explains the modern concept of total quality control and the often quoted saying that quality is everybody's business. It introduces two very important considerations. Firstly, it must be appreciated that while all sections and levels have an important part to play and a contribution to make, it is essential not to neglect the specific responsibilities and specialized techniques involved. If everyone is responsible there is a danger of no-one being responsible. The administration of an effective quality control system requires clear direction, a co-ordinated policy and efficient training in the accepted methods, and this will probably mean, particularly in large organizations, a quality control department, headed by a well chosen and well qualified chief. On the other hand, there is a danger of taking far too restricted a view and thinking that good quality control will be achieved simply by the setting up of such a department and defining its functions and activities.

If one has a dirty or untidy factory, poor supervision, low morale or inefficient design and engineering, good quality control is impossible. The best modern scientific techniques will be useless unless accepted and applied by all concerned down to the primary producer. The best results will only be obtained if the organization as a whole becomes quality minded, and, with the right lead and encouragement from top management, understands the objective and trusts the tools provided for its achievement. Someone must be responsible for promoting quality control schemes and for advice and training in statistical techniques, but success will only follow if line management gives conscious encouragement to quality control work, is prepared to accept what major changes may be involved and to foster activities and stimulate interest among those who in the end will produce and achieve the desired result, the right quality at minimum cost.

2 Inspection and Quality Control

> 'The main opportunity for improved economics of quality conformance lies, in most companies, through defect prevention.'
>
> J. M. Juran
> *Quality Control Handbook*

IN CARRYING OUT his responsibility of seeing that goods will give satisfaction to the customer, the manufacturer uses the techniques of inspection and quality control, and it must be appreciated that there is an important distinction between them. Inspection is the examination of the finished or partly finished product to prevent material which will not give satisfaction or does not comply with specification reaching the customer or passing on to the next stage of manufacture. Quality control has a much wider meaning, and comprises the techniques which are used to eliminate the causes of defective production and to ensure that the product, or at all events as high a percentage of the output as possible, will comply with requirements. In quality control one examines and studies all stages of the process: the design or formulation, the raw material, and through intermediates to the finished product, with the object of avoiding the necessity for rejection later. Quality cannot be inspected into a product: the inspector, as such, can only pass, reject or return the product for retreatment. Thus the work of the inspector can be regarded as somewhat negative.

Although the inspector cannot improve quality directly, the information he produces, if properly classified, presented and fed back to the manufacturing stage, can be a vital factor in quality control. The more closely inspection and production are integrated, the more effective such feed-back of information can be. It will be useful to consider how it came about that in so much of industry, production and inspection have become functions which are widely separated.

Before the industrial revolution, the craftsman satisfied himself that his product was fit for sale, and corrected or touched up such defects as he thought necessary. No doubt he examined his apprentice's work, rejected it or made him correct his mistakes, and instructed him how to do a better job next time, until he was satisfied that the young man was sufficiently reliable, but beyond that no separate inspection function existed. The introduction of power-operated machinery, and the emphasis on speed of output which came with the industrial revolution, resulted in the operative having little time to inspect his or her work. To protect the firm's reputation and avoid loss of sale through dissatisfaction of the customer, the manufacturer found it necessary to introduce inspection as a separate function. The inspection frequently took place days or even weeks later than production and often in a separate room and under separate supervision.

Particularly with the introduction of piece-rate, a situation was created where two sets of operatives were being paid for mutually conflicting objectives. The Lancashire cotton industry is a typical example of such a development. The weaver's earnings depended on the number of pieces of cloth she got from her looms. Time spent correcting faults meant lost wages, an incentive on chancing imperfect work. As a safeguard against such neglect, fines were introduced for faulty work. The inevitable effect was increased conflict of interest and a widening of the gulf between the producer and the inspector.

In these more enlightened days when piece-rates have been replaced by premium bonus and fines by quality incentives the position is little different. Failure to earn the expected level of quality bonus is regarded as being penalized and resented little less than a fine. There is a strong disincentive for a producing operative to give any help to the inspector by passing forward information which will help him in his task of fault-finding.

Managers who have risen from the shop floor, or who keep good contacts with their shop stewards and operatives, will be only too familiar with the practices which go on, sometimes with the connivance of foremen, as a result of this conflict of interests.

Ways and means are devised of tricking the inspector, borderline material may be deflected to an inspector who is known to be lenient or unobservant of a certain type of fault, reject material from one day's output may be reintroduced on the line and great satisfaction expressed if the inspector fails to spot what he rejected on the previous day.

In modern quality control an important feature is to overcome this conflict of interest between the producer and inspector and to integrate as closely as possible production and inspection. The more quickly and effectively information can be fed back to the producer, enabling him to take corrective action, the more perfect material will be produced. Under these conditions, the inspector becomes a valued member of the production team, helping not only to maintain quality, but incidentally to increase output. One of the greatest causes of lost output is time spent correcting faults, and if action can be taken in time to prevent faults from arising, many of the consequent stoppages of production can be avoided. There is often a strong case for the production manager being in charge of inspection of his own product. This helps him appreciate and accept his responsibility for the end product and gives him a strong vested interest in ensuring that the quality of his product is maintained. The logical extension of this principle is where possible to make the producer his own inspector, and so to return in some measure to the advantages enjoyed by the craftsman who saw that his own work was up to standard. This is not always possible, but the following case study illustrates the principles involved.

At a factory making wallcoverings in the USA, the conventional system of production and inspection as separate functions had been followed for many years. The wallcovering was printed, finished and trimmed in the manufacturing department and batched up in jumbo rolls. It was then transferred to the inspection department, where, under separate supervision, it was examined for faults—usually two or three days after printing—checked against colour and design standards, cut up into conventional 12 yard lengths and faulty material downgraded or scrapped. The workmanship, by ordinary standards, was good and normal yields ranged between 90 and 95%. It was decided to integrate inspection with production in the following way. Printing, finishing and trimming were incorporated in tandem and the machine operated by two men, who were both qualified printers. One controlled the printing machine while the other, who was provided with all the viewing facilities necessary, was responsible for batching into jumbo rolls. The two men changed jobs at half time on the eight-hour shift. The man who was working as batcher and inspector marked any faulty work on the selvedge with a tab. The jumbo roll was then sent to an automatic machine, which cut and rolled it into twelve yard lengths. All pieces free of tabs on the edge were then packed for despatch as perfect material. The tabbed pieces were passed to a reject inspector, who decided whether they should be graded for sale as seconds or scrapped. It was soon found that the percentage of perfect material rose to approximately 97% and, what was perhaps even more important, customer complaints were reduced. After running the new scheme for some time, the manager responsible experimented with more frequent changes between the two men during the shift, and found it an advantage to reduce the spells on printing and inspection to less than four hours. The efficiency of a visual inspector is known to fall off rapidly with time, particularly if the proportion of sub-standard material is small. The reject inspector's work was so light that one man could cope with the output of printing machines working three shifts and have time to spare for other duties connected with labelling and despatch. After giving a wage advance to the labour concerned, the saving in manpower and in rejection costs made a very useful addition to the company's profits. The important features in this case study were that in the new arrangement:

(1) any deterioration in standard of quality was communicated to the machine operator within a matter of minutes and corrective action taken quickly;
(2) the system encouraged pride in good workmanship. Operator and inspector understood each other's job perfectly and worked together as partners.

It should be noted that the men who co-operated in the success of the changed method were rewarded by an advance in their regular weekly wage. It will require little thought to realize that the scheme would have been quite impracticable with the payment of a quality bonus based on the percentage yield.

Our first case study concerned a process where the essential features were appearance, colour, design and the absence of local defects. We shall now take an example of a well integrated quality control scheme where we are concerned with properties which can be measured rather than judged.

At a factory in Canada manufacturing vinyl tiles, the essential features required if the product is to meet the customer's requirements are:

(1) linear dimensions and thickness to be within specified limits;

(2) dimensional stability. The tiles must not alter in size after cutting, to an extent which would cause trouble in fitting;

(3) flexibility and resistance to indentation must be satisfactory for the purpose intended, and comply with appropriate specifications;

(4) the shade of the tile must be correct to standard.

The tiles must also be free from objectionable local defects, but in this case study we shall confine our attention to the four items listed above.

All the tests were carried out alongside the production plant by a technician working on each of the three shifts. Samples were taken from the production line every 10 minutes for testing of thickness, linear dimensions and shade, every 20 minutes for hardness and flexibility, and every hour for dimensional stability. Information was fed back to the foreman, who was able to take corrective action if any property was in danger of getting outside the control limits. We shall discuss in Chapter 5 how the foreman, or machine operator, is enabled to decide in a case of this kind, on the basis of the measurements taken, whether to adjust his process or leave well alone. Meanwhile we shall confine our attention to the important feature of speed of feed-back, in the absence of which we are bound to be faced with the production of substantial quantities of defective material before corrective action can be taken. There was no difficulty regarding this aspect in the measurements of thickness, linear dimensions, hardness and flexibility, all of which tests could be completed and charted within a matter of minutes. Successive samples were also compared to give warning of any colour drift away from the approved standard. The problem lay in dimensional stability, for which the Canadian Government specification laid down that the material should pass a test which required a minimum of 6 hours to carry out. Failure on this test would, therefore, mean the rejection of a substantial amount of material. The difficulty was surmounted by the technical staff devising a test which could be completed in 20 minutes and satisfying themselves by a comparison of results that the factory test was reliable, i.e. if the material passed the factory test it would also safely pass the Government standard test. It will be appreciated that as samples were taken each hour, and testing required 20 minutes, there could still be the danger of rejection of something approaching $1\frac{1}{2}$ hours' production if a failure was detected. To meet this difficulty warning limits were laid down well inside the acceptance limits, so that when the warning signal was given that dimensional stability was approaching the danger limit, corrective action could be taken in time to avoid the necessity for actual rejection.

In this way a true quality control system was operated, not to reject material which did not comply with specifications, but to ensure that it was made right in the first place. The quality control technician, far from being regarded as an outsider and a critic, became a valued member of the production team, helping them to turn out the quality which the customer demanded, to make corrections when they became necessary, in good time, and to avoid unnecessary stoppages.

Inspection can be carried out in two ways: 100% inspection in which every article is tested or examined, or inspection by sampling. Where inspection or testing involves destruction, as for instance in the testing of electric bulbs for length of life, sampling inspection is the only possible method. In Chapter 4 we shall discuss the elementary principles of sampling theory, but here we shall deal with a common misconception, that 100% inspection is, where possible, the most reliable method. It has been established as a result of extensive research that no inspector is ever perfect; when examining material for defects he will always miss a certain proportion; efficiencies of 80–90% detection are common with good inspectors and efficiencies over 90% are rare.

In the Cavendish Laboratory in the early 1930's, Lord Rutherford used to organize a course of training for research students before they were allocated to their particular lines of investigation. One item on this course was an aptitude test of counting the flashes of α-particles on a screen. The flashes were completely random in time and position on the screen, and while some students were better than others on the test, no one ever recorded the full 100% or agreed exactly with another student watching the same screen. Here the observer was looking for one thing only—a flash of light. The inspector, who has to keep an eye open for a variety of defects on material moving over a table or conveyor, has a much more difficult task. Studies have shown that the results obtained by different inspectors, however experienced, differ widely. In particular, the inspector is much more likely to spot the faults he is expecting and to miss the unusual.

The following series of tests is reasonably typical. Two experienced inspectors examined five batches each of 1600 linoleum tiles, and removed those which they considered defective. Each batch was then mixed up, the rejected tiles being identified by a mark on the back. A second pair of experienced inspectors then inspected the same tiles and made their rejections. The table on page 13 gives a summary of the results obtained.

These results show how variable subjective judgement can be in visual inspection.

If therefore production contains defects, subjection to 100% visual inspection is no guarantee that no defect will reach the customer.

The only certain method of ensuring that no defective material

Table 2.1

Trial No.	Team No. 1 rejects	%	Team No. 2 rejects	%	Common rejects	Rejected by Team 1, passed by Team 2	Passed by Team 1, rejected by Team 2
1	3	0·19	17	1·06	2	1	15
2	30	1·88	13	0·81	7	23	6
3	66	4·14	37	2·32	31	35	6
4	61	3·82	41	2·57	31	30	10
5	36	2·25	35	2·19	14	22	21
	196		143		85	111	58

reaches a customer is to make the material right in the first place. Where this cannot be guaranteed, the next best method is for each process involved to observe and report forward to the final inspection the defects noted. This is equivalent to two or more 100% inspections, none of which is itself perfect, diminishing but not removing the probability that some faults will get through.

Such a method was in fact introduced in the manufacture of linoleum tiles, subsequent to the study described above. Inspection points, with good viewing facilities, were established following the intermediate stages of manufacture. Each roll when it reached the final stage of cutting into tiles was thus provided with a report sheet grading it as either, (a) fit for cutting and packing throughout except where clearly marked, (b) faulty or inferior throughout, (c) requiring 100% inspection for local faults. In this case the report form showed the nature of the fault which the final inspectors should watch for and its location.

It soon emerged that the proportion of batches which could be classified as (a) was normally about 90%, and the remainder of the batches came in class (c). The virtual elimination of class (b) resulted from the fact that each process or shift, being responsible for reporting forward on the quality of its production, quickly spotted and corrected any continuous fault which developed. This system not only improved the yield and reduced customer complaints to virtually negligible proportions, but resulted in a useful saving in labour, through the big reduction in the numbers required on what had previously been a laborious final inspection on the whole production.

We have thus seen that the essence of modern quality control is to develop an integrated scheme for production and inspection, with the object of producing the article we require correctly, as distinct from making it and sorting good production from bad.

Such a scheme must be designed so that producer and inspector work together as partners with the combined object of giving the customer what he requires at the price he is prepared to pay, and there should be nothing in the scheme which is likely to encourage conflict of interest or antagonism between them.

It is useful to classify defects according to their seriousness, and to gear quality control and inspection procedures accordingly. This is illustrated by the following classification based on a Ministry of Supply scheme.

(1) *Critical*—The passing of any such fault is likely to result in serious consequences, e.g. danger of loss of life or serious injury, or the virtual certainty of a major complaint and damage to the supplier's reputation.

(2) *Serious*—(i) Probability of the goods failing to give standard performance. (ii) Likelihood of complaint if the proportion so defective is appreciable.

(3) *Less serious*—(i) May cause some failure in use. (ii) Likely to cause the customer some trouble, e.g. in installation or maintenance. (iii) Major defects in appearance or finish, which however have no bearing on performance or utility.

(4) *Minor*—(i) Minor defects in appearance and finish. (ii) Technical faults which, while undesirable, are unlikely to cause trouble or poor performance.

The following figures are quoted from an article dealing with the above classification as typical acceptable quality levels (AQL) which might be tolerated with a classification of this kind.*

* *The Machinist*, April (1952), p. 577.

Class of defect	Average percentage not to exceed
(1)	Nil
(2)	0·25%
(3)	1%
(4)	3%

It will be appreciated from what has been said above that a single 100% inspection, particularly if it is of a visual nature, is no guarantee that critical defects will not reach the customer. If possible the process and product should be designed and the standard of housekeeping, good order and training be such that critical defects do not arise. In the food industry, for example, it has been said that it is more important to inspect the factory than to inspect the product. Where the danger of critical defects arising cannot be guaranteed it is essential that co-ordination of quality control and testing throughout the process and the feed-forward and feed-back of information regarding such faults is of a very high standard.

3 Some Human Aspects

'It is the individual, his latent skills and talents, which are the unseen assets which do not appear on the balance sheet of a company and which in the end dominate its performance.'

Rivett and Ackoff
A Manager's Guide to OR

IN OUR LAST CHAPTER we saw how important it is that conflict of interest and outlook between producers and inspectors should be avoided if the best results are to be obtained. In considering the human aspects of a sound quality control scheme, it will be well worth our while to consider in a little more detail the very real problem of how to make an incentive scheme aimed at obtaining maximum productivity compatible with the correct attitude to the quality and reliability of the product. In the USA, where mechanization and automation are more highly advanced, where the shop floor worker is much more financially minded and where in consequence productivity per man is considerably higher than in Great Britain, direct financial incentives are much less used in productive industry. It must be recognized, however, that a large part of British industry has not reached that stage of evolution where output incentives could safely be dropped without a serious effect on productivity.

The first essential if an output scheme is to be co-ordinated effectively with good quality control is that management thoroughly appreciates and accepts its responsibilities with regard to both output and quality; the second is that the primary producer must be brought fully into the picture. Let us deal with these matters in turn. F. W. Taylor, usually regarded as the father of work study, stressed in his classical book *Scientific*

Management[2] four prime responsibilities of the production manager: selection of the best method, selection of the worker, training the worker in the selected method and planning the correct day's task for the worker. Provided these responsibilities are accepted and carried out, it is perfectly possible to have an effective output incentive combined with satisfactory quality control. Neglect any one of them and the result will be a failure. If an unsuitable or badly trained operative is paid to obtain maximum output it is inevitable that faulty material will be produced, or if the day's work is not planned in accordance with what a worker trained in the correct method can do without over-exertion, there will again be the temptation to produce without the necessary attention to quality. The great danger lies in piece-rate or other incentive schemes, in which rates have been fixed without adequate study, and where operatives are left to choose their own method of working without proper training. I have yet to meet the man or woman who, given the right training and understanding the purpose of the work, does not prefer to do a good job rather than a bad one. If management sets up a scheme which, through lack of planning and training, encourages output with the neglect of quality, it is failing in its responsibilities.

Where on the other hand the necessary study, planning and training has been done, it will, in general, be found that the correct method of working results is good output combined with satisfactory quality. Under such conditions the main cause of lost output is the correction of errors, and as a well integrated quality control scheme minimizes the danger and frequency of having to correct such errors, the quality technician becomes the producer's most valued ally, the man who helps him to obtain maximum output and to earn maximum bonus. Superimposition of a quality bonus, or of quality sanctions, is then not only unnecessary but undesirable. Thus, while recognizing that the ideal for combined high productivity and a satisfactory and reliable product is a high day rate combined with a high degree of efficient mechanization or automation, we see that it is by no means impossible to combine sound quality control with a well designed output incentive scheme. The essential factor is that the manager must not evade his responsibilities and expect a financial incentive to produce the results required in quality and output without a great deal of careful thought, planning and training, both by him and by those he controls.

Let us now turn to our second point, the importance of the operative and how to bring him fully into the picture and obtain the best co-operation from him in producing good results. We have seen that quality bonuses are in general an ineffective method of obtaining good work, and that this is particularly true if the amount of bonus earned depends on an inspection process separated from production in time and space. The perfect out-turn depends on a number of factors, some of which are under the operative's control and many of which are not. If that operative's pay fluctuates from week to week in accordance with perfect yield, it will usually be found that a fall in the level of earnings, far from proving an incentive to do better next week, results in annoyance and frustration. This applies particularly if the output and quality bonuses are calculated independently. The output bonus is a thing the worker understands and can directly influence, but he may be far from clear as to how he can influence the quality earnings. His natural tendency is therefore to concentrate on output earnings, accept the quality bonus when it is good and grumble about it when it is poor. In other schemes the quality factor is intimately linked with output, the two commonest forms being (a) piece-rate where only perfect output qualifies, or (b) where output bonus is increased by a factor if yield exceeds a certain percentage. While better than the independent output–quality incentive, such schemes are still open to the objection that they cause conflict of interest between producer and inspector, and if the factors which determine yield are not understood or controllable by the operator, can again result in annoyance and frustration.

If then fluctuating financial 'incentives' are ineffective, what other steps can be taken to encourage operative participation in obtaining a good yield or a good quality product? The fundamental factors are training in the correct method, understanding of the objective and pride of workmanship. Each of the factors can be encouraged and developed. While condemning fluctuating quality incentives or sanctions, let us first look at two forms of monetary reward which in no way conflict with our objective.

The first of these is rewards for suggestions and ideas which improve quality and reliability, or which save cost without endangering or lowering quality standards. Such suggestions should be encouraged and rewards should be realistic in value. A man or woman who contributes something effective and lasting to the improvement of a company's products or processes deserves a reward, and such a reward, unlike a fluctuating bonus, increases rather than diminishes pride in the job. A second form of financial recognition which can often be given with good effect is an advance on the *rate for the job* when a certain level of training is passed or level of skill reached. Unlike the conventional quality bonus, such an advance is not withdrawn when there is a temporary fall in yield, often caused by bad luck or outside circumstances. It should only be withdrawn after an adequate warning and proper study of the reasons for deterioration in results. Such withdrawals should rarely or never be necessary; it is usually better if skill falls off and cannot be corrected to transfer the operative to another job. This type of financial recognition for skill and good work is much easier to apply where work is individual or in very small groups than where larger teams are necessary, because of the problem which arises whenever a new worker has to be introduced into the team. Where it can be applied, once again if properly and fairly administered it can enhance pride in the job.

Financial considerations, however, are not the only, or even the principal, factors in encouraging and obtaining the best workmanship. First and most fundamental is good training. We have already stressed management's responsibility in selecting the best method and training the operative in that method. From the worker's point of view, individual and well thought out instruction not only puts him in a position to do a satisfactory job, but creates the vital factors of interest and feeling of recognition. Training should include not only technical instruction, but explanation of the why and wherefore. This is particularly important where the product is some unspectacular mass-produced component. Let the operative see and understand what that component has to do and how it can affect the quality and performance of the finished product and there will be much more interest in producing it in the form and to the specification needed. Thirdly, training must include explanation and understanding of the necessary controls. When we discuss control methods in Chapter 5, we shall see how important it is for the operative to understand and trust these methods and not to regard them as management gimmicks.

Pride of workmanship can also be encouraged and interest enhanced by seeing the finished article and its performance. Some companies have, with good effect, made a practice of sending a workman along with the representative when a customer complaint is being dealt with, but it is also helpful occasionally to see the results of good work well done. There is no doubt that the ancient custom of shipyard workers seeing their ship launched well repays the time spent.

Another factor well worthy of consideration is making use of the natural interest in competition. Results obtained on different machines, or in different departments, compared and clearly presented can do a lot to create healthy rivalry in trying to do better than one's fellows. Where this is not possible, competition to achieve a target can again stimulate interest. Targets should be realistic, demand effort, but not be incapable of realization.

It is worth spending time and thought on how best to present data, so that they can easily be understood and appreciated. Lists of figures put up on a notice board arouse little interest. The manager or technologist who produces the information must be careful not to judge others, whose educational background may be far different from his own, too much by himself. The graphs which the works manager finds useful on his office wall may have little impact in the workroom. This is illustrated in Figures 3.1 and 3.2, which show the same data in conventional graphical form and specially prepared for shop floor presentation. A little psychology is necessary in finding what will strike home and what will pass over people's heads. In a 1966 quality and reliability campaign in a North of England floorcovering factory, when data on the costs of certain faults were published in terms of thousands of pounds per year it was found that very little

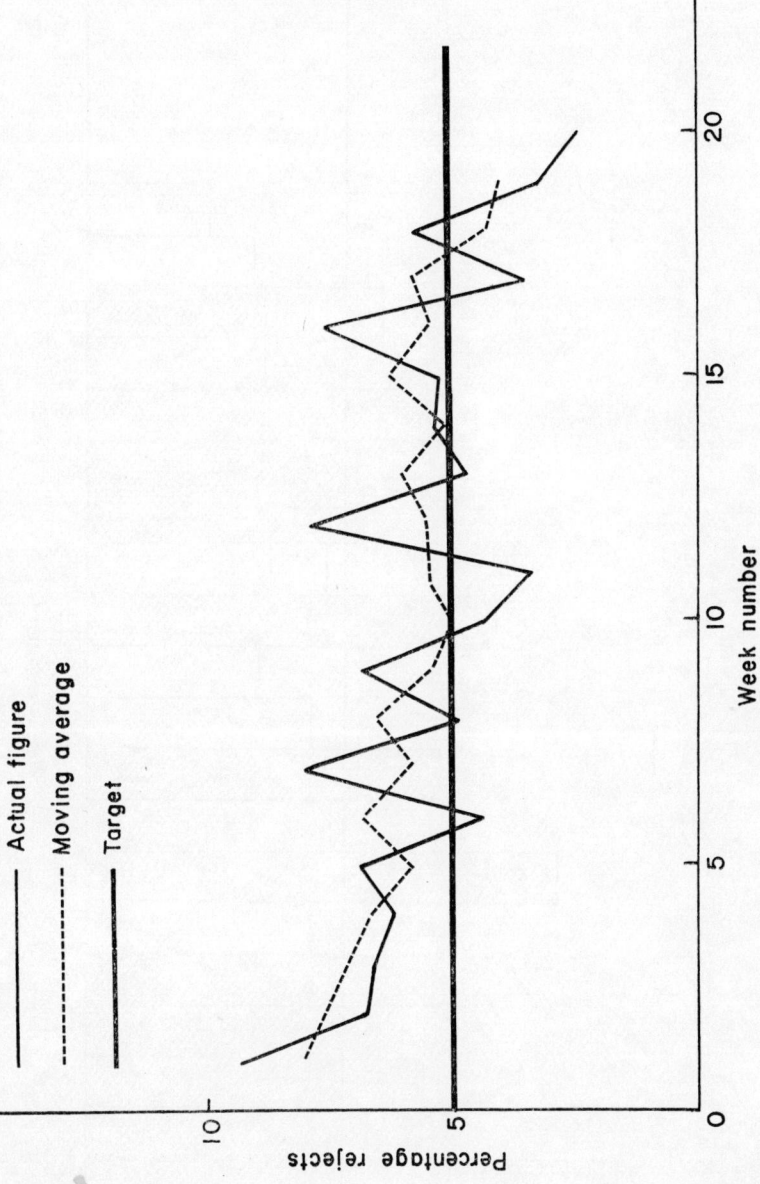

Figure 3.1. Department A. Weekly percentage rejects. Manager's graph.

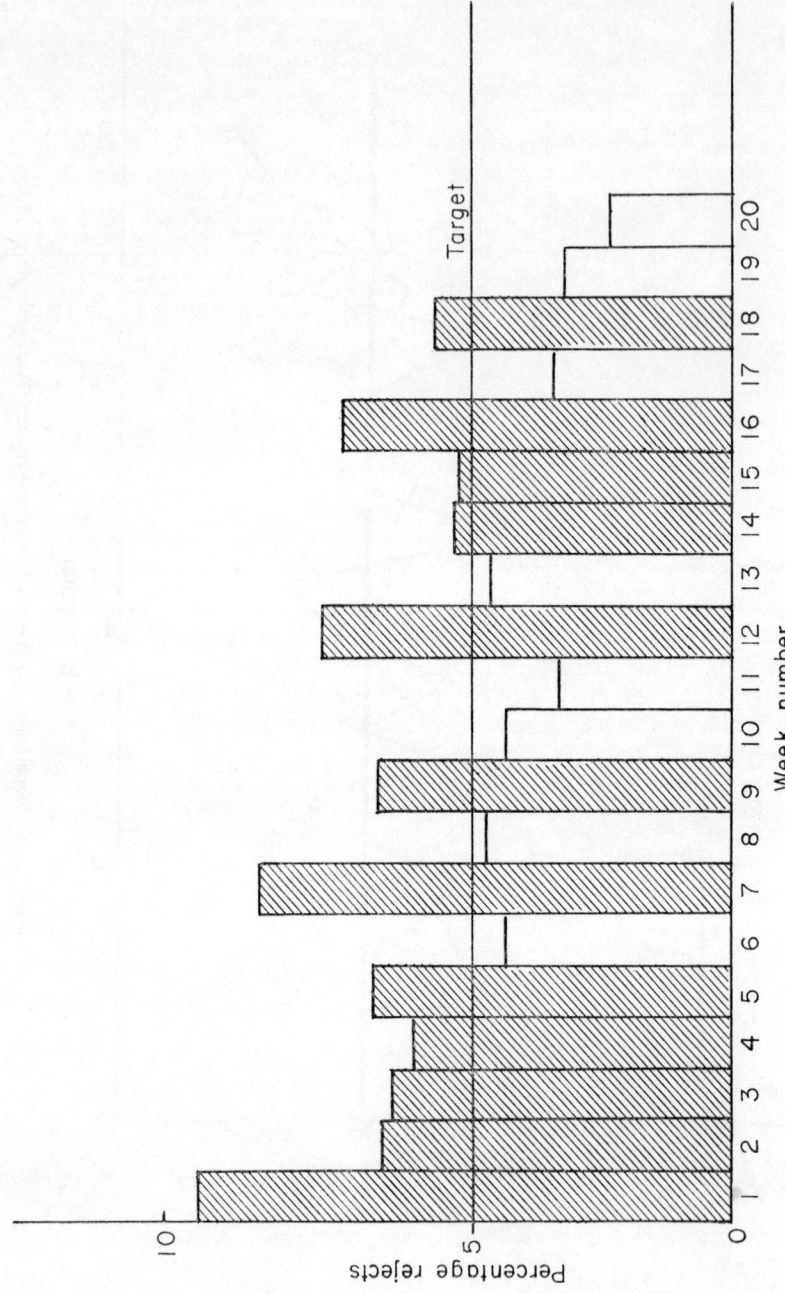

Figure 3.2. Department A. Weekly percentage rejects. Shop floor presentation.

interest was aroused. When they were expressed in terms of shillings per roll of linoleum or pounds per week they had immediate effect. They were now in terms the worker could appreciate and was used to meeting in his way of life. What applies in North Lancashire, however, may not apply in the South of England or in the USA. The important thing is to study your public and put over data in the way which will have most appeal and arouse the most interest.

Finally it must be stressed that a positive approach commending good work or outstanding achievements may be far more encouraging than continual harping about the bad. When results are below standard or customer complaints serious, it is important that proper enquiries and reports should be made, but while these are necessary, progress depends even more on recognizing improvements and abnormally good performances. It is these that do more to enhance pride of workmanship and pride of performance, which are so important in determining a company's progress and reputation. Looking forward may be much more effective than looking backward, and there is a great deal to be said for spending rather more time in studying the possibilities of future achievements than in explaining the failures of the past.

4 Statistical Concepts

'A very little consideration shows that there is scarcely a hole or a corner of modern life which could not find some application, however simple, for statistical theory and show a profit as a result.'

Moroney
Facts from Figures

WE SAW in our introductory chapter that the fundamental requirement of a product is to give satisfaction to the customer in quality of design and quality of conformance. In Chapter 6 we shall discuss in more detail some of the aspects of design for quality. Meanwhile let us assume that the article or component we are considering has been well designed and when produced to specification fulfils its functions adequately. What are the principal reasons why in practice material is frequently rejected on inspection or results in failures and complaints when it reaches the customer? Sometimes the cause is a simple failure to follow instructions, but by far the commonest cause of lack of conformance is variability. In a fraction of the production, a critical dimension may be slightly outside the limits necessary for efficient function or fitting, the strength of part of the material may be a little below standard, or minor blemishes on a proportion of the output have risen to a level which could cause dissatisfaction. It will generally be found in a well-managed factory that far more material is rejected because variation from standard spreads beyond acceptable limits than because of clear and definable errors.

It is important to recognize that some variability is inevitable and that there is no such thing under practical conditions as perfect precision. This variability arises in a number of ways: men, materials and machines are all liable to vary from time to time, or from place to place. It is no good shutting our eyes to this variability or treating it as a nuisance which should not exist; if we are to get good results we must appreciate it, measure it and control it.

Many of the properties on which the function and saleability of a product depend are measurable, for example, linear dimensions, weights, tensile strength, and electrical properties such as resistance and capacity. In such cases the actual value of the property concerned can be measured and recorded and we can introduce a scheme of control or inspection by measured variables. In other cases the factors determining the defect are difficult or impossible to measure, but the defects can be counted. Examples are slubs or smashes in cotton textiles, metal components affected by cracks or blisters, doctor scratches or smudges in printed goods, and foreign bodies or foreign colour in plastics or linoleum. In such cases the factor causing the fault is termed an attribute, and systems of control or inspection by attributes can be set up.

With the object of saving time and cost it is very frequent practice to use a system of inspection by attributes, even though the variable we are concerned with is measurable.

The most well-known example of this is rejection of articles or components for size or weight by means of go no-no gauges. As we shall see later, measured variables give far more information than simple counting of attributes and this apparent economy can prove to be a false one.

A third type of defect is one which cannot be either measured or counted, or for which it is not at all clear to see a possible method of measurement. Examples are colour or shade of goods, smoothness of finish, or degree of gloss. Sometimes, by using a little ingenuity, methods can be devised for measuring such properties and as a result quality control made much more effective. This will be illustrated by a case study on pages 41-43.

Where variables can be measured or attributes counted, statistical methods are essential to any good quality control system. This is because statistics is the branch of mathematics which enables us to handle variable quantities.

It is not still not uncommon for otherwise intelligent or responsible persons to say they distrust statistics and to shy away from the use of statistical methods. This is partly a defensive attitude due to ignorance of the techniques, and partly experience in seeing the result of 'statistics' mishandled or misinterpreted. The attitude is rather like a man who has suffered at the hands of a quack saying that he distrusts doctors. Let us then be quite clear and recognize the fact that statistical methods are an essential part of any soundly devised quality and reliability system wherever specifications can be expressed in quantitative form. The commonly held fallacy that the usefulness of statistical quality control is limited to mass production of engineering components is quite untrue. While it is in engineering that statistical quality control had its earliest and greatest impact, many process industries, including textiles, plastics, paper, chemicals, floorcoverings, rubber and glass, are using it with great effect.

A variety of statistical techniques are used in quality control, most of which are quite simple to understand and apply, and for the benefit of those readers who have not had any statistical training it is hoped that the following explanation of some of the fundamental principles will be useful.

The statistician regards any series of variable measurements, e.g. the weights of components coming off an assembly line, as represented by a mean or average figure superimposed by a degree of random variability about this mean. If in this example it is required to keep within certain specified limits of weight, two things must be watched: any drift of the mean from the central or target figure, or any excessive variation about this mean. The degree of random variation about the mean is most conveniently measured by what is known as the standard deviation from the mean. The standard deviation is the root mean square of all deviations from the mean and is usually denoted by the Greek letter σ. This is an extremely useful quantity which enables us to assess, with any degree of confidence necessary, what proportion of our output will lie between any given limits. The practical significance of the standard deviation can be easily explained. If we are concerned, as is frequently the case in quality control, with measurements which are almost equally likely to vary about our mean or target figure in either direction, the distribution of results is described as 'normal'. The standard deviation enables us to estimate the proportion of measurements in a normal distribution in accordance with the following rule:

Approximately two-thirds of all measurements lie within one standard deviation from the mean.
The great majority (approximately 95%) lie within two standard deviations from the mean.
Virtually all (approximately 99·8%) lie within three standard deviations from the mean.

The following example will make this principle clear to the non-statistical reader.

The following figures were obtained for the weights of fifty components coming off a production line, the target weight being 5·00 g.

Table 4.1

4·94	5·08	5·04	5·06	4·98	4·90	5·00	5·02	5·00	4·90
4·86	4·92	4·88	4·82	4·92	5·00	4·82	5·02	5·00	5·00
5·06	4·94	5·26	5·00	4·84	4·98	4·76	4·96	5·00	4·94
4·76	5·02	5·08	4·80	5·02	4·96	5·02	4·80	5·04	5·04
5·06	5·06	5·10	5·18	5·10	4·98	5·14	5·04	5·08	5·12

These figures are represented effectively by a normal distribution with a mean of 5·00 and a standard deviation of 0·10. If the reader analyses these figures he will find:
36 readings lying between 4·90 and 5·10 (72% within 1 σ of the mean)
47 readings lying between 4·80 and 4·90 (94% within 2 σ of the mean)
All readings lying between 4·70 and 5·30 (100% within 3 σ of the mean)
which agree well with the theoretical figures given above. Such figures are said to be in statistical control.

It is this principle which lies behind the theory of sampling and the Shewhart control chart, which are fundamental in statistical quality control when we are concerned with measured variables.

It is also important to appreciate what a statistician means by significance. If a small sample is taken from a production line and a property measured which differs from the target value at which we are aiming it is necessary to know whether this difference is likely to be due to purely chance causes, or whether it means that the process is out of control and requires adjustment. Again, if we are testing an alternative supply of raw material or examining a laboratory trial aimed at quality improvement, can we be satisfied that the tests and measurements we have made are sufficient to establish the superiority of the new material or suggested process alteration? The statistician deals with these questions by posing what is known as the null hypothesis. Assume that the difference from normal standards is due to pure chance; what are the odds against this being true? The following scale is generally recognized by statisticians on a wide range of problems for measuring significance.

Chance against truth of null hypothesis	Significance
Under 20–1	Not significant
Over 20–1	Probably significant
Over 100–1	Significant
Over 1000–1	Highly significant

If, therefore, we are dealing with a process which corresponds reasonably to a normal distribution, a single sample differing from the mean by more than 2σ is probably significant, warning us to be careful and test again, whereas a sample differing from the mean by 3σ is significant and demands action. Differences appreciably greater than 3σ, or two successive readings close to 3σ, clearly take us into the highly significant range. In more complex cases, significance tests are applied to determine the odds against the null hypothesis. These tests, which the reader will find fully described and explained in textbooks of statistics, enable us to answer such questions as the following:

(1) In comparing the proportion of defects made in two alternative production methods, can we be satisfied that method A gives significantly better results than method B? (The chi squared test for proportions.)

(2) In comparing the means of test results on physical properties of alternative supplies of materials, can we be

assured that one is significantly better than the other, or do we need further data? (The t test for comparing means.)

(3) Is the degree of variability significantly higher using procedure A than using procedure B? (The F test for variance.)

The reader will readily appreciate that an understanding of significance is essential if we are to base judgement and take action on the testing of samples.

We may sum up the principal statistical techniques used in quality control as follows:

(1) Frequency distributions
(2) Sampling techniques
(3) Control charts
(4) Statistical tests for reliability

We shall deal in this chapter with simple examples of the use of frequency distributions and discuss the fundamental principles of sampling. Chapter 5 will be devoted to control charts, and reliability tests dealt with briefly in Chapter 6. References given at the end of the book will guide the reader who needs to study these techniques in more detail or to obtain information on special or more advanced statistical methods which fall outside the simple classification given above.

Frequency distributions

The simplest method of recording a frequency distribution is the tally chart, which can be illustrated by the figures in Table 4.1. The spread in this table ranged from 4·76 to 5·26. Figure 4.1 shows the frequency in each interval of 0·05 in tally chart form. It is convenient, in order to facilitate counting up, to arrange the tally in groups of five, the fifth stroke being an oblique one through the preceding four. We see at once that the distribution is reasonably symmetrical with a strong central tendency in the group between 5·00 and 5·04. Provided the process remains in control, the more readings recorded, the more smooth and symmetrical the chart will become. We have already seen that the figures approximate to a normal distribution, and below the tally chart in Figure 4.1 the normal dis-

30   A MANAGER'S GUIDE TO QUALITY AND RELIABILITY

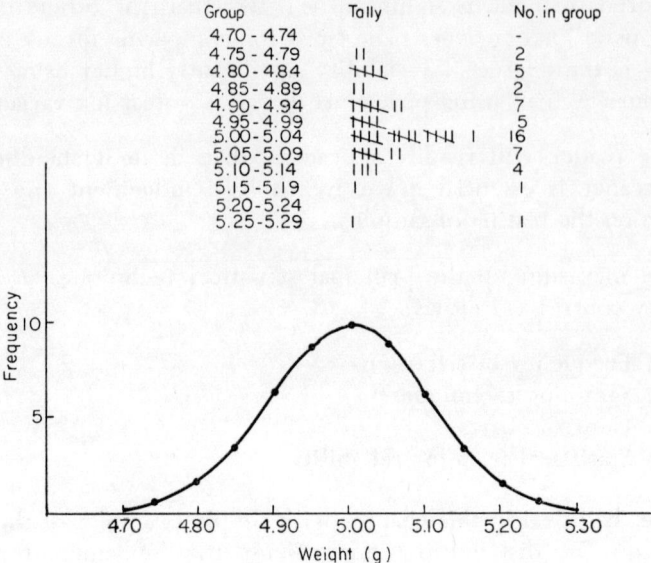

Figure 4.1.   Component weights: normal distribution.

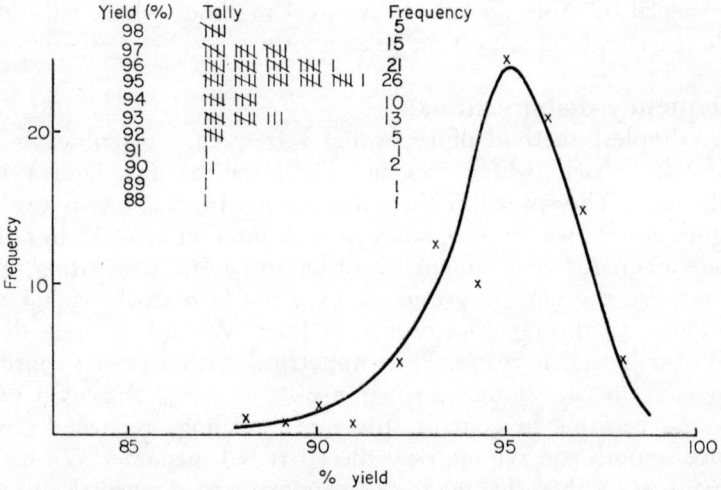

Figure 4.2.   Percentage yield: skew distribution.

STATISTICAL CONCEPTS 31

tribution which corresponds most closely to the tally is shown. Often, of course, distributions are not symmetrical, the tendency to spread being greater in one direction than the other. Figure 4.2 shows a tally chart and frequency distribution for the percentage yield of a chemical process. Here the central tendency or mode is 95%, but there is a much wider spread of results below this figure than above.

Another convenient method of recording the frequency of a series of variable measurements is a histogram, where the number of readings between each pair of limits is represented by a column proportional in height to that number, as shown in Figures 4.6 and 4.7. The following are some simple practical examples of the use of frequency distributions in quality control.

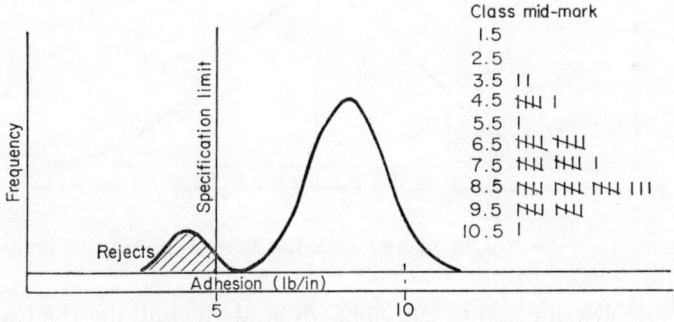

Figure 4.3. Two-headed frequency distribution indicating rejects due to a specific error.

(1) On a certain process roughly 10% of the production was failing to pass an adhesion test, where a minimum separation force of 5 lb per inch was specified. Was the cause of defective production some specific fault on a proportion of the manufacture, or is the general spread of normal production such that under present conditions this percentage of defective material is to be expected? A frequency distribution of the type shown in Figure 4.3 indicates a specific fault in workmanship, which should be relatively easy to locate and avoid. On the other hand, if we obtain a distribution of the type shown in Figure 4.4, it is clear that the defective production is due to

inherent variability in the process, and that its elimination will require an entirely different approach. It is no good hoping to obtain better results by tightening up supervision or blaming the operative; we must either overhaul the process and raise the mean adhesion say from 8 to 10, shifting the

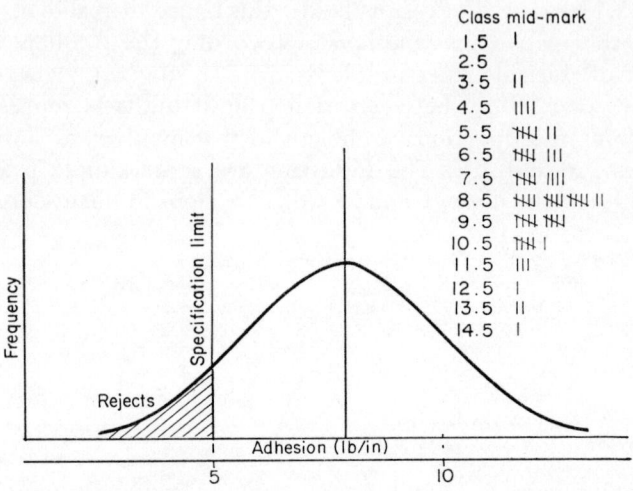

Figure 4.4. Normal distribution spreading beyond specification limits.

whole distribution to the right, or study technically the causes of the wide spread and take steps to reduce the inherent variability.

(2) Is process machinery capable of producing the product consistently to designer's specification? A part is required with a diameter of 5.00 ± 0.05 mm. Samples are taken from a test run and measured. (See Figure 4.5.) A distribution of type A is satisfactory and shows there should be no difficulty in meeting the tolerances. Type B shows that inherent variability is too great and that production of defective material is inevitable. If a satisfactory product and yield is to be obtained we must therefore overhaul our method or alter our machine to cut down variability, or obtain agreement for a wider tolerance. A third possibility is a distribution of type C. Here all the test samples complied with specifi-

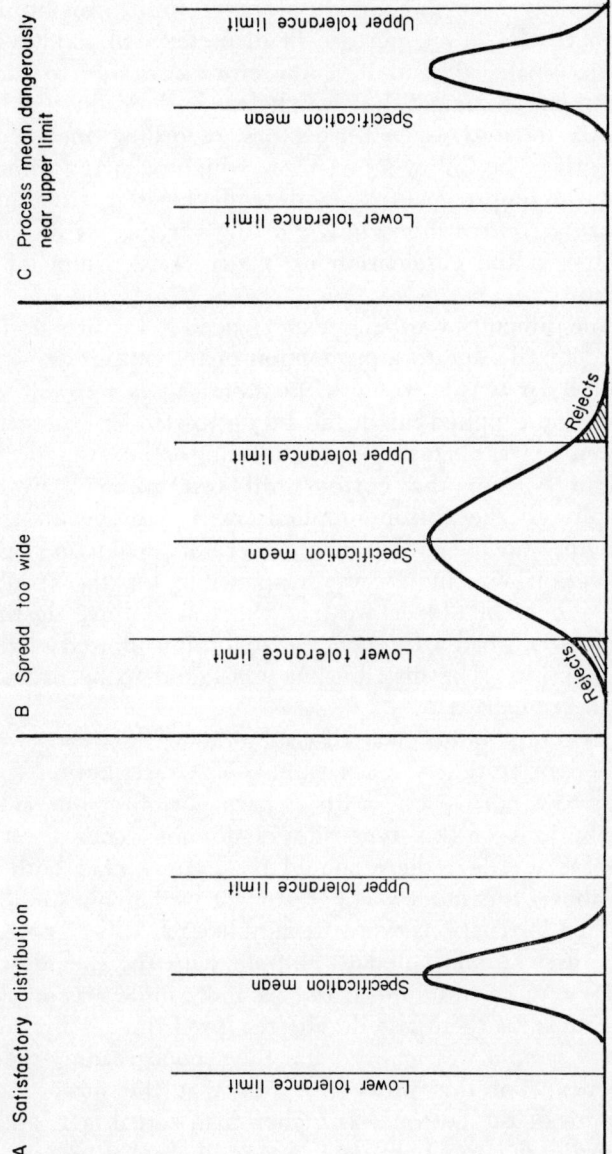

Figure 4-5. Capability study distributions.

cation and it might be assumed that the position would be satisfactory. We see, however, that the central tendency is dangerously near the upper tolerance limit. Any tool wear which results in an increase in diameter will soon result in oversize material, and it is therefore advisable to make an adjustment to correct the mean before bulk production begins. (3) Are operatives or inspectors recording measurements correctly? The following example which occurred some years ago in a company which was introducing statistical methods of quality control shows how plotting a frequency distribution resulted in the elimination of incorrect recording of information.

Some difficulty was being experienced in the production of a coated textile due to a proportion of the output being below standard for tensile strength. Possible causes were either that the cotton supplied might not be up to standard or that some weakening was occurring in the finishing and coating processes.

Figures from the cotton mill test room showed that virtually all the cotton supplied was up to specification for strength, and the mill manager therefore maintained that his production was in no way responsible for the trouble. A quality control officer was set the task of investigating the problem and had a frequency distribution plotted of the test room results. The distribution was found to be of the form shown in Figure 4.6.

It was now quite clear that something was wrong with the method of testing and recording. The distribution approximates to a normal one with a sharp cut-off at one side, and distributions of this type simply do not occur in strength tests for textiles; there should be a fair spread both below and above the mode. The tester on being questioned said that her instructions were to sample each roll of cotton and if the first sample failed to comply with the specification to ask for a second sample. If the second sample was satisfactory the batch was then passed. The reading from the first unsatisfactory sample was ignored and the second reading recorded. The result of this procedure was that the mean recorded strength of the cotton was higher than actual fact, and as by the ordinary laws of chance it was unlikely that two successive

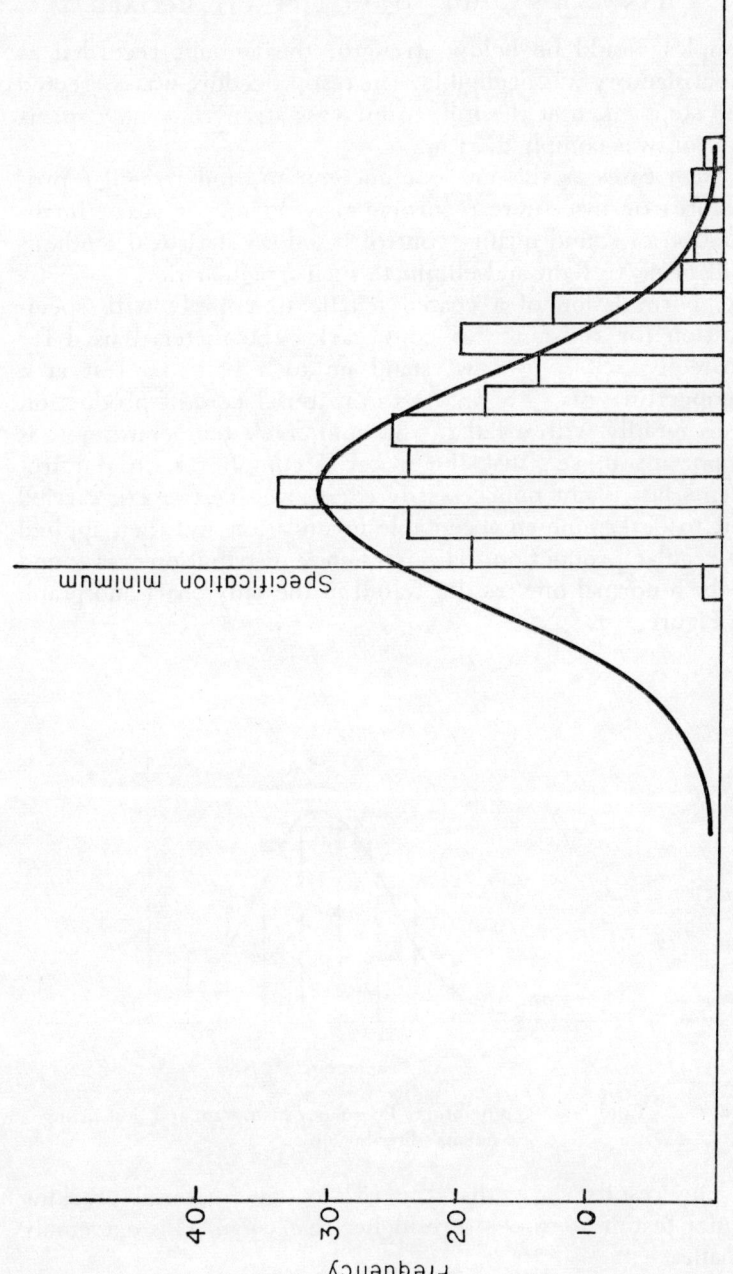

Figure 4.6. Irregular distribution with cut-off at specification minimum.

samples would be below strength, the amount recorded as unsatisfactory was negligible. The test procedure was corrected and steps taken at the mill to increase strength so that specification was complied with.

Such cases as this are not uncommon, and irregular procedures or inaccurate recording may go on for years. Introduction of sound quality control based on statistical methods will bring to light and eliminate such irregularities.

(4) Formulation of a coated textile to comply with specification for resistance to cold-crack. The material used for protective clothing must stand up to a cracking test at a temperature of $-20°$ C. As the material cost of production rises rapidly with a fall in the cold-crack temperature, it is important to see that the material complies with requirements but is not unnecessarily expensive. Tests were carried out to determine an acceptable formulation and then applied to regular production. The frequency distribution was found to be a normal one, as illustrated in the tally chart and graph in Figure 4.7.

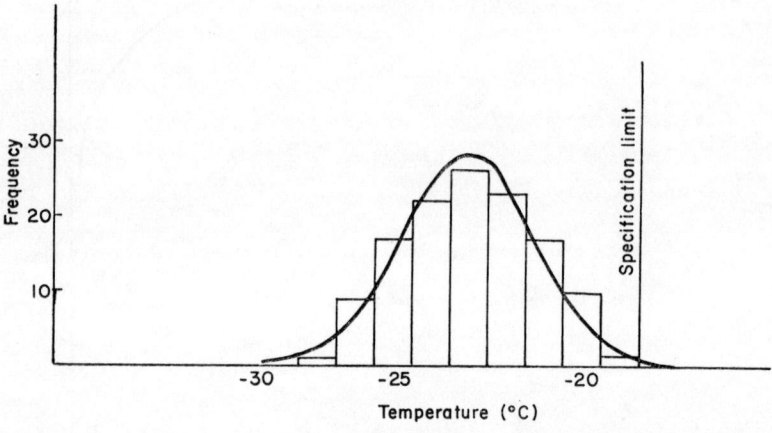

Figure 4.7. Cold crack temperature. Frequency histogram and best fitting normal distribution.

The results show that the risk of any material cracking under test at a temperature higher than $-20°$ C is extremely small.

These examples, all based on actual factory experience, show how useful the study of frequency distributions can be in quality control. We have seen how they can help to narrow the field of investigation when trouble-shooting, avoid the cost of faulty work through inadequate tests, help to spot and eliminate bad practices in testing and recording and avoid either defective work or unnecessarily costly formulations. They are, in fact, a most useful tool for the quality control technician and can be applied to a wide variety of problems.

Sampling techniques

While it is outside the scope of this book to deal in detail with the theory of sampling, it is important that we should understand the fundamental principles on which good sampling depends and appreciate how misleading and dangerous results can arise if sampling schemes are not properly designed. Inspection can be either 100%, in which every article purchased or produced is tested, or every yard of fabric viewed and examined for faults, or by sampling in which case batches are accepted, returned or rejected on the basis of examination of a proportion of the lot under consideration. Frequently sampling is the only possible method because the test is destructive, as in the case of testing electric bulbs for length of life or the testing of cotton or other fabrics for tensile or tearing strength. Even when the test is not destructive, sampling inspection is often used to avoid the expense of 100% inspection. In particular, goods inward inspection is usually done by sampling methods.

The first principle in sampling is that a single sample gives no information about variability, and, in the absence of knowledge of such variability, is useless. Single sampling is therefore only effective where our knowledge of the process and the methods of control is sufficient to ensure a high degree of homogeneity in each batch examined. Some processes in the chemical industry are of this nature; provided the specified procedure with regard to time and temperature has been carried out, each batch is uniform within fine limits, and all we have to watch out for is an error in composition, due for instance to wrong weighing. Under such conditions single samples give the information we require. On the other hand, in a far wider range of industry variability occurs within the batch. This applies to engineering production in general, textiles, paper, and a wide variety of other materials and products. Furthermore, in the case of goods inward inspection from outside suppliers, we have often no assurance that individual deliveries represent individual manufacturing batches. Yet how often in industry, when such conditions obtain, do senior executives say, 'Bring me a sample', or even 'Bring me a representative sample', and make an important decision regarding purchase or change of process on this basis?

The second important principle is that samples must be taken in a random manner, that is, in a way which avoids any likelihood that the sample will be biased or different from the general distribution of the batch. For instance, it would be quite wrong, if we require a sample of 25 from say 5000 articles passing over a conveyor belt, to take all from an adjacent area; to obtain a representative or random sample we should work on a time basis, e.g. if the production time is 50 minutes, samples are taken at two minute intervals. If we require a sample of 10 from a 60-bag lorry load of raw material, a good method would be to take a sample from every sixth bag unloaded. Each case should be considered on its merits, the essential point being to think out a scheme which eliminates as far as possible the probability that the sample will be affected by some special factor which is not applicable to the batch as a whole.

Thirdly, it is important to appreciate that when we are sampling by attributes and deciding whether to accept or reject a batch on the basis of the number of defectives in a sample, it is more economical and effective to take large samples from large batches than small samples from small batches. For instance, much more dependable results would be obtained by examining a random sample of 100 from a batch of 10,000 than from one of 10 out of 1000, whereas a sample of 1 from each batch of 100 would be useless, for the reasons given above when we were discussing the dangers of single sampling of variable goods. The unsoundness and ineffectiveness of many common sampling acceptance schemes, in which small samples are taken from relatively small lots, is illustrated by the following example quoted from E. L. Grant's book *Statistical Quality Control*.[3] In this example the sampling scheme consisted of taking a sample of 5 from each lot of 50, accepting the lot if the sample contained no defectives, and rejecting it if the sample contained one or more defectives. In all, 50,000 articles were submitted in 1000 lots and 5000, five from each lot, inspected. The average percentage defective throughout the complete delivery was 4%. Grant shows that on the ordinary laws of probability 815 lots would be accepted and 185 rejected, and the accepted lots would contain 3·6% defective. Hence as a result of all the work put into sampling and inspection and the rejection of 18·5% of the delivery, the purchaser only effected a negligible improvement in quality.

Finally, it is important to appreciate that much larger samples are required in inspection by attributes than in inspection by measured variables. This last principle can be well illustrated by consideration of the figures given in Table 4.1.

Let us suppose that these figures were obtained from the examination of a sample of 50 components selected in an unbiased or random manner from a production batch of 5000 and our specification demands that at least 90% of the production must lie within the limits 4·80 and 5·20. Examination of our sample has enabled us to determine the general pattern of distribution of weights in the batch as a whole and to know with a very high degree of confidence that the production is well within our specification. On the other hand, let us suppose that the sample had been examined not by recording actual weights (i.e. by measured variables), but by simple counting of defectives (i.e. by attributes). It can easily be shown that if the actual proportion of defectives in the full batch was in fact 5%, there would have been approximately a 10% chance of finding at least five defectives in the sample, and an 8% chance that the sample would be entirely free from defectives. These figures show that a sample of 50 (1% of the batch) would have been a sound basis for passing or rejecting the batch if we were inspecting on the basis of measured variables, but quite inadequate on the basis of inspection by attributes.

We have thus seen that in inspection by attributes, a single sample is valueless and a small sample, such as the 5 out of 50 quoted from Professor Grant's book, or even the 50 out of 5000 discussed above, very unreliable. There are, of course, problems involved in working on the basis of large samples from very large batches. The first problem is obtaining a sample which is truly 'random' or representative. If necessary this can be overcome by using a stratified sample, i.e. by dividing the full batch into sub-lots, and taking sub-samples from each sub-lot proportional to its size. The second problem is the expense and delay involved in the rejection or turning over to 100% inspection of the occasional very large batch. This can be met by devising rules for repeat sampling when the first sample indicates danger, or by sequential sampling. We shall return to the subject of sampling inspection in Chapter 7 when discussing the economics of a modern quality control system. The reader who requires further information in sampling theory and procedures should refer to the books listed in the Bibliography on pages 122 and 123.

When goods inward inspection is done by sampling methods it will be appreciated from what has been said above that many existing schemes are far from being reliable or satisfactory. It will be realized that the appropriate sampling scheme depends very much on the degree of variability which can be expected. Whereas a simple sampling scheme may be satisfactory in the case of a supplier whose record and reputation for consistency is good, it may be difficult or impossible to devise any sampling scheme to cover the case of the supplier whose goods are extremely variable. For this reason many companies today are adopting the principle that different suppliers should be given what is known as a vendor rating, a measure of their degree of reliability. If this is good, responsibility is placed on them to supply goods in accordance with specification, and goods inward inspection is dispensed with entirely. Suppliers with poor vendor ratings are dropped. This can be illustrated by the following case quoted from E. L. Grant's book, *Statistical Quality Control*.[3]

'The Niagara Frontier Division of the Bell Aircraft Corporation analyzed acceptance and rejection records covering nearly 35 million parts purchased from 458 companies. Only 1·95%

of these were found defective and rejected. But it was found that:

    277 companies supplied parts   0 to   1·99% defective
     39 companies supplied parts   2 to   4·99% defective
     31 companies supplied parts   5 to   9·99% defective
     44 companies supplied parts 10 to 19·99% defective
     36 companies supplied parts 20 to 49·99% defective
     31 companies supplied parts 50 to 100%   defective

In a case of this kind, the unreliable suppliers can be eliminated from the list, and goods inward inspection and its attendant costs avoided.'

The following case study reproduced from the author's book, *Scientific Method in Production Management*,[4] illustrates many of the points we have been discussing, particularly the advantages of using measured variables rather than attributes, the devising of a sound sampling procedure and the importance of vendor rating.

One of the principal raw materials in manufacture of linoleum is wood flour, and it is important that the shade of wood flour should be controlled, as variation in shade can cause serious variation in the depth of a light or pastel shade in the finished product. It was established practice for the laboratory to examine a sample from every delivery of wood flour and grade it for colour. The most reliable wood flour was imported, and considerably more expensive than the home-produced supplies. The light-coloured imported wood flour was used for the lighter and more delicate shades. It was found in practice that while some deliveries of home-produced wood flour gave a good shade on the sample, they proved unreliable in practice, and their use had to be avoided, in spite of a satisfactory laboratory report.

The vital step that enabled progress to be made was the development by one of the laboratory technicians of a photometer, which measured the degree of whiteness of the wood flour. Instead of a visual opinion, grading it as light, medium or dark, it now became possible to express in terms of a number, the degree of light reflectance, a measure of its whiteness. Thus, a poor coloured wood flour might be 70, medium 80 to 85, and 100 would represent the highest possible standard—the whiter than white of the detergent manufacturers. It now became possible to use statistical methods, which had been impossible when the colour was judged in qualitative terms, and to devise a reliable acceptance sampling scheme based on statistical sampling theory, which would give a picture for each delivery of the range of shade included. Plotting the results obtained on a curve showed that there was a wide variety in shade both in imported and home-produced wood flour, with a considerable overlap (see Figure 4.8). Instead of taking a single sample, a random sample of fifteen was taken from each delivery. The fifteen samples were blended in threes and a photometer reading taken from each of the five combined samples.

Thus, it became possible to arrive at a dependable figure for the minimum whiteness of each batch. As a result it was found that a large proportion of the home-produced wood flour was up to standard, and that some suppliers were much better for consistency of shade than others. Suppliers now could be kept up to the mark, given a vendor rating for reliability, and the consistently unreliable eliminated. It was soon found that it was unnecessary to purchase any imported wood flour, and the consistency of shade noticeably improved. The saving in cost by not having to import was substantial. A final logical step in a case of this kind is, having established reliability on the part of suppliers, to simplify the acceptance sampling procedure or, if the suppliers' record is sufficiently good, to discontinue it entirely.

We may sum up this chapter as follows. Variability is usually the principal cause of defective work. It cannot be eliminated, but can be controlled. Such control is greatly facilitated by measuring our variables. Statistics is the branch of mathematics which enables us to handle variable figures and is thus essential to any sound quality control scheme where possibility of measurement exists. Sampling techniques used in quality control must be devised on sound statistical principles, failing which misleading and dangerous results can arise.

The reader will appreciate from this brief introduction to sampling techniques how important it is that sampling schemes

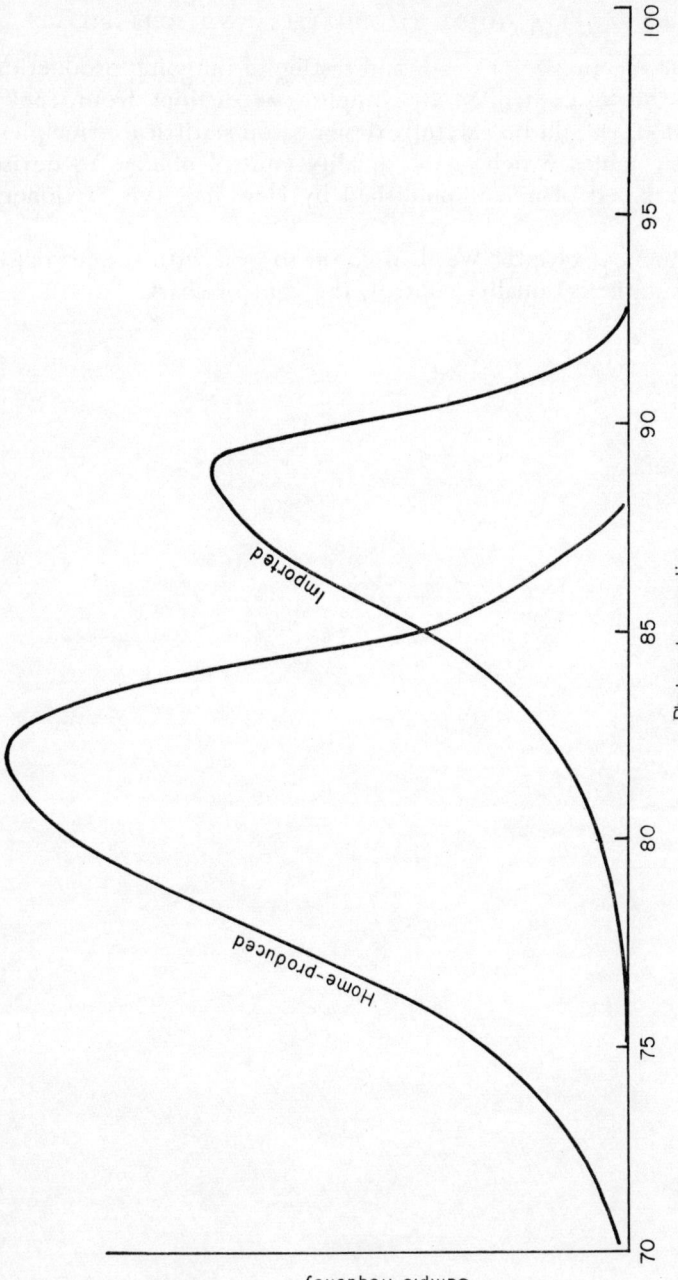

Figure 4.8. Whiteness of home-produced and imported wood flour. Reproduced from the author's book *Scientific Method in Production Management* by permission of Oxford University Press.

both for acceptance of goods and testing of outgoing production where this is controlled by sampling as distinct from 100% inspection, should be carefully designed on statistical principles. Sampling tables which enable quality control officers to devise soundly based plans are published by Her Majesty's Stationery Office.[5]

In our next chapter we shall go on to deal with the principal tool of statistical quality control, the control chart.

5 Control Charts

'Some stable system of chance causes is inherent in any particular system of production and inspection. Variation within this stable pattern is inevitable. The reasons for variation outside this stable pattern may be discovered and corrected.'

E. L. Grant
Statistical Quality Control

THE BEST-KNOWN and probably the most important tool in statistical quality control is the control chart, introduced by W. A. Shewhart during the 1930's. Control chart technique is based on the recognition that measured quality of a manufactured article is always subject to a certain degree of random variation as a result of chance causes. This variation arises partly from lack of precision in the process itself and partly from lack of precision in the means of measurement available. Variation of this character is inevitable. The proper use of the control chart enables us to discover and take action to correct variation outside the normal pattern. Expressed in another way, any series of measurements on the dimensions, weights or physical properties of our product can be regarded as made up of a long-term trend, the movement of the mean, superimposed by inherent or random variation. The Shewhart control chart enables us to separate the long-term trend from the random variation and also to detect when the degree of random variation begins to exceed normal proportions. It enables us to see and take action when action is needed and at the same time see when it is best to leave well alone, to concentrate our efforts on the tracking down and correction of real production troubles and to avoid wasting time on unnecessary and fruitless investigations.

To illustrate this let us consider the following example. Figures for percentage yield of a paper plant for three successive weeks were 81·2, 77·5 and 73·9. Should the manager responsible initiate an investigation and report on the reasons for the fall in yield? Records showed that the mean yield for the previous 12 months was 75% with a standard deviation of 3·3%. Now we saw in our last chapter that with a normal distribution 95% of all readings can be expected to lie within two standard deviations from the mean, i.e. between 68·4% and 81·6%. The three readings in question all lie within these limits and the variation thus represents less than a 1 in 20 chance that there is any irregularity. Investigation is therefore unnecessary, unless the downward trend continues for at least another week. This conclusion is confirmed by examination of the percentage figures for the nine-week period in the centre of which the three weeks quoted occurred.

The figures were: 69·0, 72·5, 75·7, 81·2, 77·5, 73·9, 76·8, 72·4, 77·6. The mean of these figures is 75·2, which is slightly above the 12-month average, and it will be clear that nothing was in fact lost by suspending judgement and not conducting an investigation.

We saw in our last chapter, when discussing sampling, that the methods differ according to whether we are dealing with measured variables or attributes. The same applies to control chart technique. Methods used for hour to hour control on the shop floor also differ from those most suited to following longer-term trends. We shall devote the rest of this chapter to describing and illustrating four types of chart used in statistical quality control.

Control chart for measured variables

If our object is to make goods of consistent quality based on a specification which can be expressed in terms of measurements, two things are necessary: to keep the mean as steady as possible and near to the central point of the specification, and to control variations about this mean. If both factors can be controlled so that measurements are kept within the specification tolerances, we shall be able to achieve our objective of consistently turning out acceptable goods. If study of the random or inherent variability shows that it exceeds the specification tolerances, then no amount of stimulating, bribing or threatening the production operative concerned will result in our obtaining a good yield, and as we shall see may possibly do more harm than good. We have only two courses of action open to us; to obtain permission to widen the designer's tolerances, or to get down to a careful study of the causes of variation and reduce them within acceptable limits. This can be illustrated by means of a simple example.

Let us suppose that we are turning out on a production line quantities of tiles, and our specification limits for thickness are $2 \cdot 00 \pm 0 \cdot 10$ mm. In order to check whether the process is in satisfactory control and that thickness conforms with specification, samples are taken every ten minutes and the thickness gauged. The first 48 samples, reading across from left to right, give the following results:

Table 5.1

2·00	1·99	1·95	1·98
2·01	2·04	2·00	2·03
1·98	2·03	2·01	1·93
2·02	2·02	2·00	2·07
2·01	1·99	1·93	2·04
1·99	2·03	2·04	2·01
1·98	1·99	2·03	2·02
2·02	1·97	2·02	2·01
1·96	2·02	2·00	1·98
2·04	1·99	1·98	2·02
2·04	1·97	2·00	1·97
2·02	2·02	2·04	2·01

The mean of these figures is $2 \cdot 002$, which is very close to the target or central point of the specification, and it will be seen that the main range is from $1 \cdot 94 - 2 \cdot 06$ mm with three recorded readings, two of $1 \cdot 93$ and one of $2 \cdot 07$, just outside this range. Expressed in statistical language, the figures correspond closely to a normal distribution, with a mean of $2 \cdot 00$ and a standard deviation of $0 \cdot 03$. The figures and hence the process are in statistical control.

If we were setting up a control chart for this process, we should draw a central or target line at $2 \cdot 00$, warning lines at

2·06 and 1·94 (two standard deviations from target) and action lines at 1·91 and 2·09 (three standard deviations from target). The control chart tells us that the process is in good control and should be left alone. It also tells us that the production, with a very high degree of probability, corresponds with specification and that no further inspection for thickness is necessary. The control chart is illustrated in Figure 5.1.

The chart tells us the limits between which we expect variability for purely chance causes to occur. Statistical theory tells us to expect that about one reading in 20 will lie outside the warning limits and about one in 500 outside the action limits. Adjustment should only be made if we get two or more successive readings outside the warning limits, or a reading outside the action limits. Such occurrences would be rare if the machine was in control and therefore indicate when they are encountered that there is good reason for adjustment.

It should be noted that the fixing of our warning and action limit lines is not based on the design specification, but on the observed inherent variability of the process. In this case the tolerance limits ($\pm 0·10$) and our action lines ($\pm 0·09$) happen to be near together, but this is not necessarily the case. If the inherent variability had been measured by a standard deviation of 0·05, and the action lines therefore at $\pm 0·15$, nothing the operative could do would have produced consistently satisfactory material.

Let us consider what might well have happened if we had no control chart and it had been left to the operator to make what adjustments he thought necessary, guided by common sense and experience.

A conscientious operator after taking the reading of 1·93 in line 5 of Table 5.1 (marked P in Figure 5.1) might adjust the centering of his machine to increase thickness by an amount of 0·07 mm. Meanwhile inherent variability in the process continues as before and instead of the next reading giving a figure of 2·04, as in our table, the next measurement is 2·04 + 0·07 or 2·11 mm. This reading is actually outside the specification limit, and the operator now makes a further adjustment of —0·11 mm in an attempt to get back to target. Inherent variability would meanwhile have resulted in a drop of 0·05, so

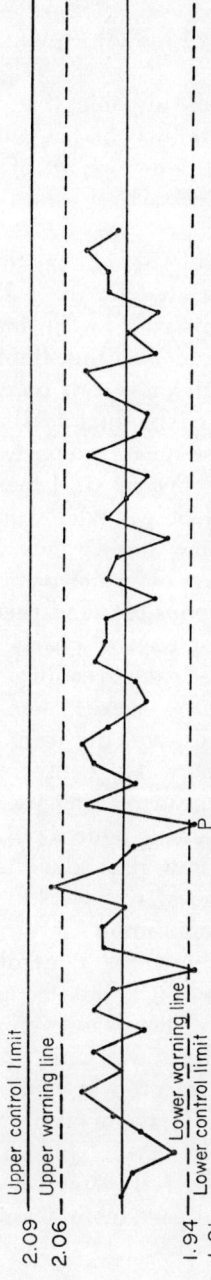

Figure 5.1. Control chart. Thickness of tiles, process in statistical control.

with the additional effect of his adjustment, the operator throws the gauge to 1·95.

Thus, far from obtaining his objective of less variable thickness, by his well-meaning efforts he has caused what is known as *hunting*, increased variability as a result of over-control, and has actually produced, unnecessarily, some material outside the specification limits.

How real this danger of introducing hunting and increasing variability is can be illustrated by the following case from the author's own experience. Prior to the introduction of control chart technique on a calender rolling linoleum sheet, two men were employed gauging thickness, one on each side of the material. Adjustment was made to the thickness controls by the calender driver on the basis of these measurements, and such adjustments were often as frequent as five or six times an hour. The newly appointed quality control officer who studied the process soon found that it was showing some signs of over-control. The first step taken was to put over the principles involved to the manager and foremen responsible, and then to the men directly concerned. Following this, control charts were introduced, and instructions laid down that gauge readings were to be made to a definite plan. Two readings were taken at each side of the machine every ten minutes, and the mean of the four measurements recorded on the chart. The centering of the machine was only to be adjusted if a mean reading was obtained outside the action limits, or two successive figures outside the same warning limit. As a result of the new procedure it was found that such adjustments once the process was working smoothly were very rare, in fact it was not uncommon for the machine to run for a full eight-hour shift without the controls being touched. As gauge readings were now much less frequent an operative was saved on each shift, and what was even more important, the accuracy of gauge was quite appreciably improved. In due course the amount of material rejected at final inspection for wrong gauge, which had previously averaged between one half and one per cent, fell virtually to zero. This made it possible when measurements had been well established to eliminate final inspection for thickness, apart from a sample from each batch for record and audit purposes.

Let us return now to our figures and suppose that the process continues and a further 48 readings are recorded as in Table 5.2.

Table 5.2

1·99	1·99	2·03	1·99
1·98	2·02	2·00	2·00
2·00	1·99	2·01	2·01
1·99	2·01	2·01	2·02
2·04	2·08	2·03	2·07
2·09	2·08	2·03	2·10
2·00	2·05	2·02	2·01
2·01	1·95	2·02	2·01
1·98	1·98	2·04	2·00
2·01	1·96	2·00	2·04
2·04	2·05	2·00	1·97
2·06	2·03	1·98	1·98

The figures are plotted on a control chart in Figure 5.2 and illustrate the use of the chart when something occurs which disturbs the centering of the machine. At the point A we obtain a reading outside the upper warning line, followed by one on the upper action limit. It is clear that something has gone wrong and that an adjustment is needed. This takes a little time and two further high readings are obtained before the process is back in control at point B. Production now continues satisfactorily until the end of the shift.

It will have been noted that in the case study on rolling linoleum sheet we did not record every reading on the control chart but the mean of four readings taken at regular intervals. Usually this procedure is better than recording each individual measurement. To illustrate the procedure, let us return to Tables 5.1 and 5.2 and treat each horizontal row as a sample of four. Table 5.3 shows the mean value for each row and the range, i.e. the difference between the highest and lowest reading in each sample. The conventional symbols for the mean and range of sample readings are $\overline{X}$ and R, so the type of charts we are now going on to consider are often referred to as $\overline{X}$ and R charts.

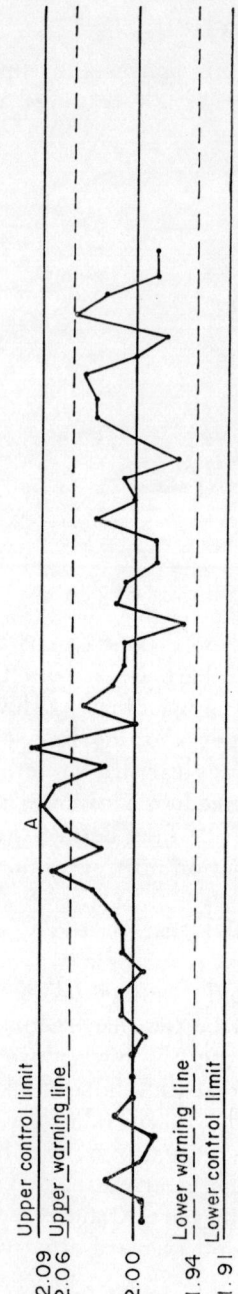

Figure 5.2. Control chart. Thickness of tiles, process out of control at A.

Table 5.3

Sample No.	Mean ($\overline{X}$)	Range (R)
1	1·98	0·05
2	2·02	0·04
3	1·9875	0·10
4	2·0275	0·07
5	1·9925	0·11
6	2·0175	0·05
7	2·005	0·05
8	2·005	0·05
9	1·99	0·06
10	2·0075	0·06
11	1·995	0·07
12	2·0025	0·03
13	2·00	0·04
14	2·00	0·04
15	2·0025	0·02
16	2·0075	0·03
17	2·055	0·04
18	2·075	0·07
19	2·020	0·05
20	1·9975	0·07
21	2·00	0·06
22	2·0025	0·08
23	2·015	0·08
24	2·0125	0·08

It will be seen that these figures of sample averages are less variable than the individual readings. In consequence, the control chart from sample averages (the $\overline{X}$ chart) requires closer limits than the chart for individual readings, and in this case, using samples of four, the warning lines should be at 2·03 and 1·97 and the action lines at 2·045 and 1·955.

It is also often useful to plot a range chart showing the figures recorded for the differences between the highest and lowest figure in our sample of four. This chart gives us additional control on our degree of variability, and can be marked with action limits in the same way as the $\overline{X}$ chart. One advantage of plotting control charts with sample averages instead of individual readings is that this has the effect of normalizing the distribution of results. This means that if the distribution of individual readings is not truly normal, for example if variations below the

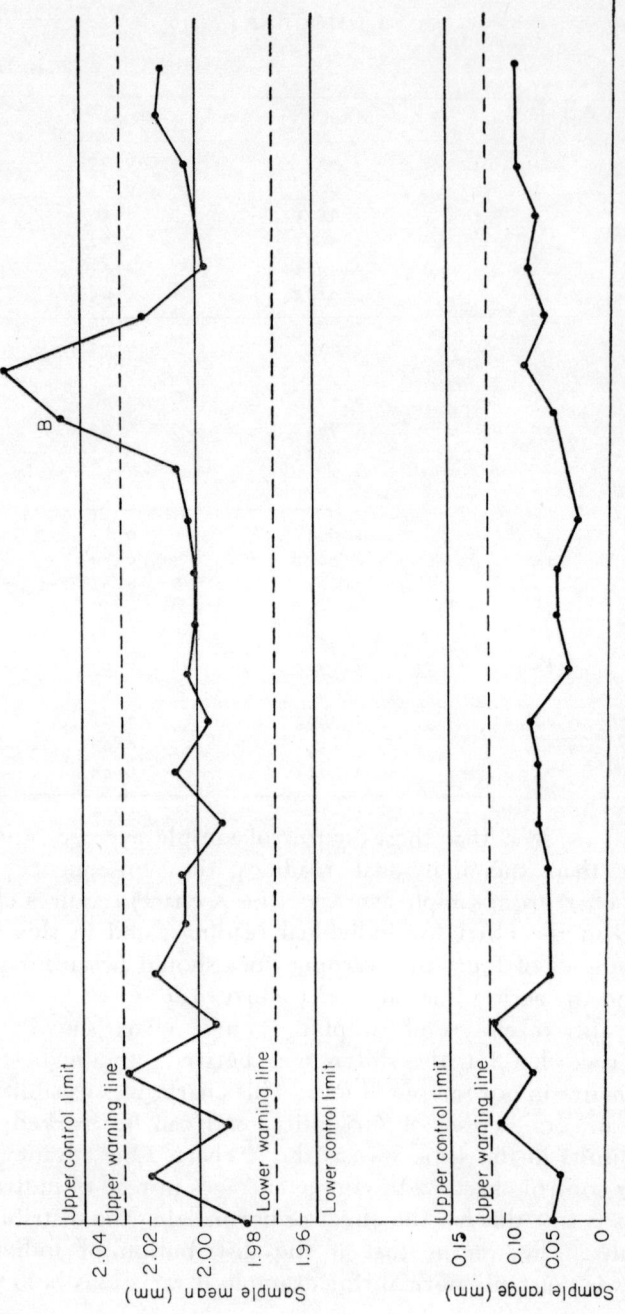

Figure 5.3. Mean ($\bar{X}$) chart and range (R) chart. Thickness of tiles. Based on figures in Table 5.3.

CONTROL CHARTS

central figure are greater or more frequent than those above, the distribution of sample averages will be much more regular and in consequence give better control. Figure 5.3 shows the mean and range charts for the figures in Table 5.3. It will be seen that the $\bar{X}$ chart shows that adjustment was required to restore correct centring at point B, but that the range chart remains in control throughout. This shows clearly that the trouble which occurred was due to a shift in machine centring and not to any increase in random variability.

It is sometimes thought that the value of control charts is limited to the case of machine-controlled repetitive production, and that they have no application to hand-made work, where it is human care and attention rather than reliability of a mechanical device which determines whether the work will be right or wrong. It is of course true that a machine well designed and maintained will probably continue to turn out material within regular limits for a considerable time, and adjustment will only be required perhaps a few times per shift. The control chart gives us warning when corrective action is required, and it will most probably be the mean which will drift and can be corrected by a simple adjustment.

Increase in variability of results, on the other hand, is probably a much more long-term consideration, brought about by gradual wear which might develop over a period of months. From the point of view of process control the mean chart is therefore frequently much more important than the range chart.

With hand-controlled operations the position is different. Here the main causes of irregularity are human. Fatigue, boredom or distraction, through noise or interruption, may result at any time in a relaxation of care and irregularity of work. The tired operative, while still turning out material which averages on the target figure, allows more variation to take place. Thus, in hand-controlled operations it is usually the range chart which is the more important. Let us illustrate this by a simple example. A factory is making a component in which the weight is important and is specified as $2 \cdot 00 \pm 0 \cdot 05$ oz. The shop has installed an automatic machine for this job, but it is unable to meet full output requirements and a number of hand-controlled machines are still in use. Output on the automatic machine is 500 compo-

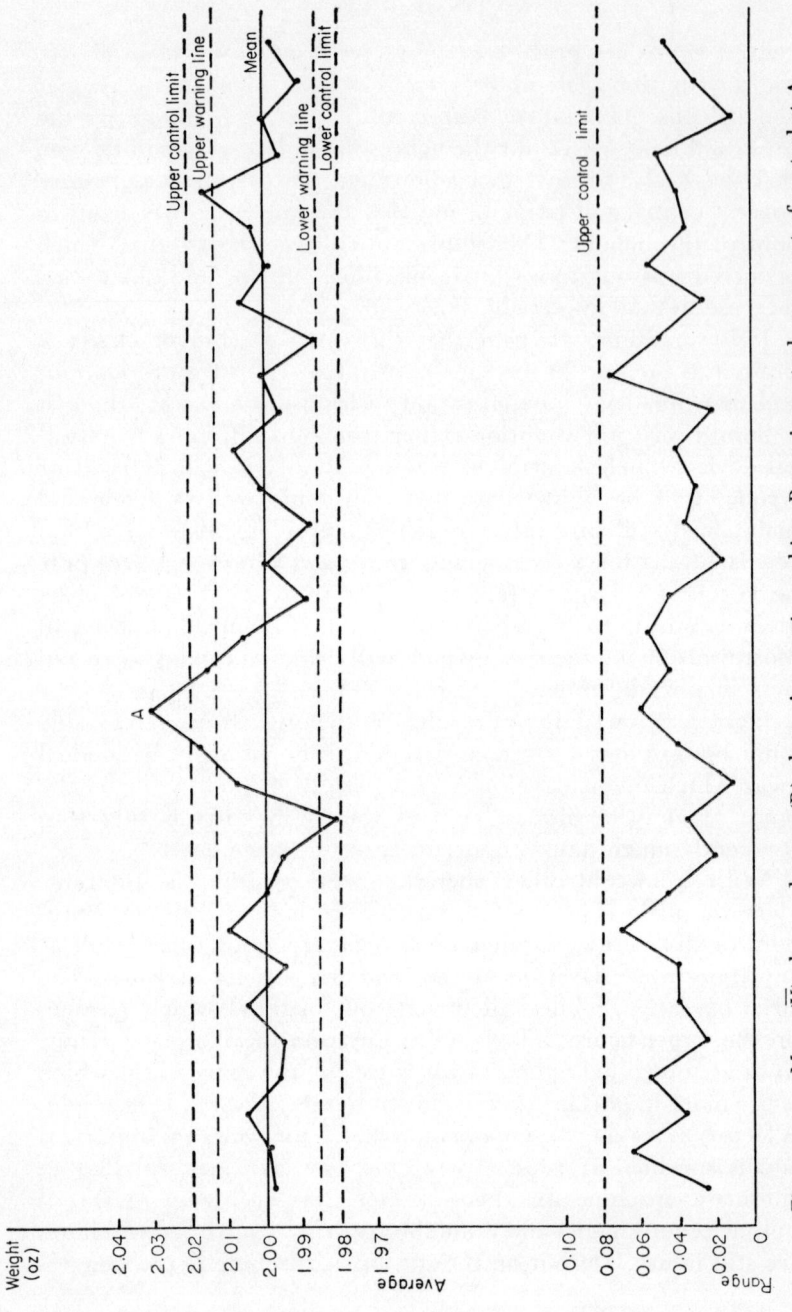

Figure 5.4. Mean ($\overline{X}$) chart and range (R) chart. Automatic machine. Range in control, mean out of control at A.

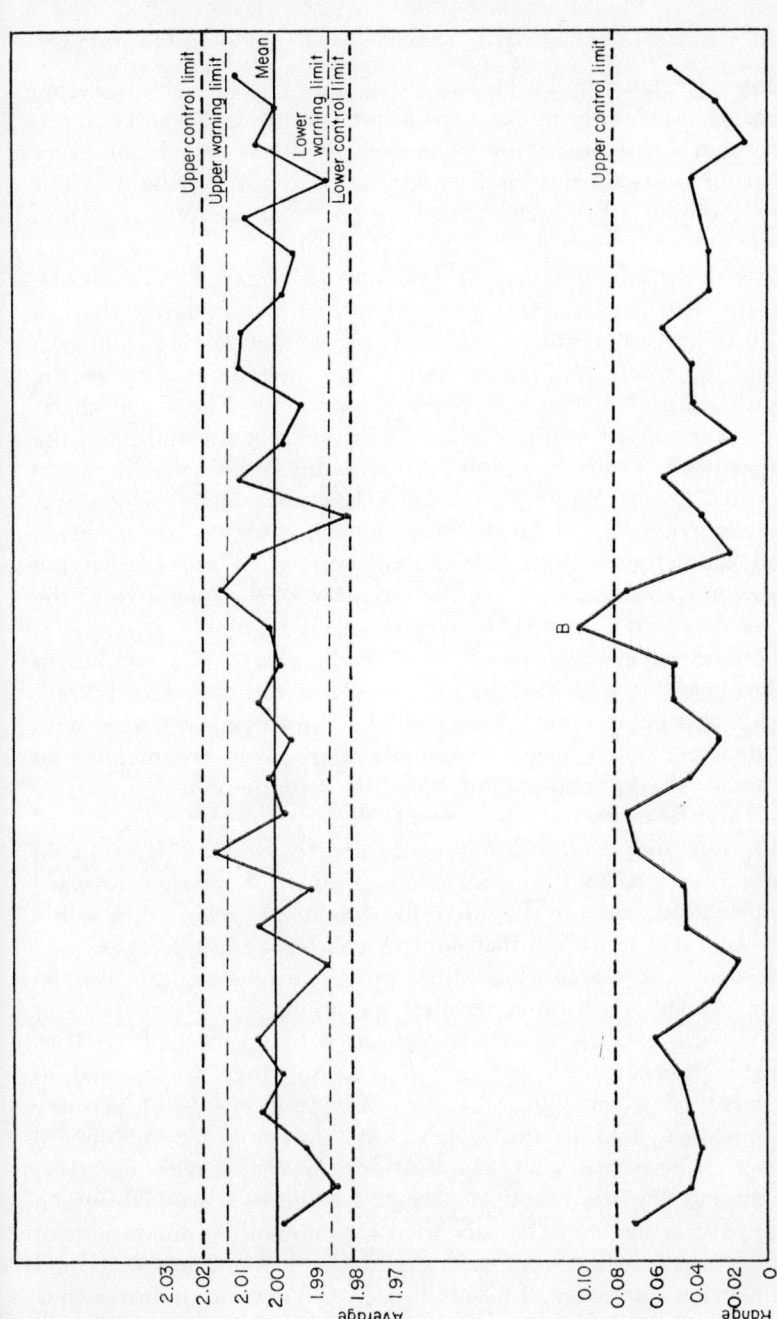

Figure 5.5. Mean ($\overline{X}$) chart and range (R) chart. Hand-operated machine. Mean in control, but range out of control at B.

nents per hour and five hand-machines capable of producing 100 per hour make up the total required output of 1000 components. A patrol inspector visits each of the six machines every quarter hour, takes a random sample of four from the previous quarter hour's production, and weighs each component. Mean and range charts are plotted for each machine. Figures 5.4 and 5.5 show the results obtained on the automatic machine and on one of the hand-machines respectively. Figure 5.4 shows that an adjustment was required and made on the automatic machine at point A, where the mean drifted off standard resulting in the production of material averaging overweight. The range chart, however, stayed within the limits throughout the shift. On the other hand, Figure 5.5 shows that at point B the range increased beyond normal limits, showing a lack of control which was quickly corrected by the operative herself as soon as her attention had been drawn to it. It could, of course, be argued that the inspector could have drawn the attention of the operative to the irregularity and that there was no need to plot a chart. It is common experience, however, that the chart has a number of advantages. It is posted in full view and the operative takes a pride in it being a good one; good and bad work is reported with regularity, and in event of trouble there is no argument as to whether the inspector pointed out the trouble or not.

Apart from its value in control, the chart is useful for record purposes and, if the components are to be used later in an assembly, provides a convenient means of passing forward information, or can be used in the investigation of possible trouble. Thus, we see that control charts, while they may have their greatest application on automatic machines, can also be very valuable on hand-controlled operations.

The use of Shewhart control charts is by no means limited to the engineering industry. In papermaking they are applied to control of weight, or of thickness where this is of primary importance, and to the weights and dimensions of cardboard tubes. There are many applications in the textile industry, including tensile strength of yarn and width of woven fabrics. In the plastics industry they are used for controlling dimensions of mouldings, thickness of calendered sheet and dimensions and dimensional stability of mouldings. In the ceramic industry they are used to control dimensions. A general application, which often gives very useful savings, is to the accuracy of weights in the filling of sacks and containers. Better control enables safety margins to be reduced and still ensure compliance with minimum specified or guaranteed weights. In the chemical industry they have been applied effectively to such matters as the control of acidity (ph) of solutions.

In this chapter we have followed the common British practice of using two sets of limits, the warning limit at 2σ and the action limit at 3σ. The usual American practice is to use charts with action limits only. This has the advantage of definiteness and simplicity; the operative knows when to make adjustment or look for trouble and he is not tempted to alter matters and perhaps induce hunting when a single reading appears in the warning zone. On the other hand, two successive readings, or several near together in the warning zone, is a very strong indication that trouble has started, and where warning lines are not used it is well for the quality control engineer or other qualified person to be in a position to interpret results and advise action when this occurs. The decision as to whether warning lines should be actually printed on the chart should be made in the light of local circumstances.

A logical extension of the control chart technique is to semi-automatic and automatic control. In semi-automatic control, the dimension, weight, or other measurement which it is required to control is recorded by an automatic instrument which prints the control chart directly. The machine operator observes the printed chart and adjusts his machine as in the case of hand control when the action limits are reached. The following case illustrates the use of semi-automatic quality control.

The process consisted of the application of a polyvinyl chloride wear layer to a printed floor covering. It was important that the thickness of the applied film should be as uniform as possible and within controlled specification limits. Thickness of application was measured by an automatic gauge and printed in the form of a control chart. An electric connection rang a bell if either the upper or lower action limit on the chart was reached and the operative then knew that he should look at the chart and make the necessary adjustment. The audible signal proved to be a great advantage, as the operative was able to devote his attention to other matters when the process was in control from the point of view of even application, and did not have to be continually watching his dial or recorder.

In fully automatic control the adjustment as well as the recording of the information is carried out without the intervention of human agency. When adjustment is required, this takes place automatically, following the same principles as hand control using the chart. It is, of course, just as important in automatic control as in hand control that hunting and over-control should be avoided. The principles are the same, but it becomes possible to introduce greater refinements than are practicable in hand control. As we have seen, in control chart technique the operator makes no adjustment unless the action limit is reached, or two or more successive readings are observed on the same side of the warning limits. It might well be the case that a small adjustment is desirable before these limits are reached, for example if three or four successive readings are inside but near one of the warning lines. In such cases it would be difficult without rapid calculation for a man to decide whether an adjustment is required and if so how much. An automatic control backed by a computer, however, can make the necessary calculations and effect the appropriate adjustment.

Well-designed automatic control can assist in obtaining consistent quality by reducing human error. On the other hand, it must be stressed that automation, particularly in assembly processes, demands very high standards of quality control in the earlier stages of the process such as the production of components. A badly fitting part which in manual assembly might have been easily set aside or adapted may, on an automatic process, cause a serious stoppage to costly plant. Maximum automation thus has a double effect, increasing the necessity for sound quality control in the subsidiary processes, but helping to ensure a consistent product by minimizing the dangers which arise through the imperfections of man's work and judgement.

We shall conclude this brief introduction to the theory and use of Shewhart control charts for measured variables by summarizing the main points which have been illustrated.

(1) The control chart is a tool of control for the benefit of the man in charge of the machine. It tells him when he should make adjustments and when to leave well alone. If it is simply used as a device for giving information to a manager about the behaviour of the process, it ceases to be a control chart.

(2) The limits on the control chart are not based on the designer's tolerances, but on the observed inherent variability of the process. If these limits are wider than the designer's tolerances the machine operative cannot be expected to produce a consistent perfect out-turn, and it is the job of management to see that the process is overhauled or the tolerances revised.

(3) It is often better to record sample averages on the chart instead of individual readings, but if this method is adopted the warning and limit lines must be moved in, in accordance with the size of the sample. (The distance of the limit lines from the centre is inversely proportional to the square root of the sample size, for example, for a sample of four, the distance of the limit lines from the centre or target is half that for individual readings.)

(4) Unnecessary or too frequent adjustment to a process or machine, even when done with the best of intentions, causes hunting and increased variability.

Control chart for attributes

Where it is not practicable to express the difference between defective and perfect material by measurements, it is frequently possible to use a control chart based on attributes. As in the case of the chart for measured variables, samples are taken from production at regular intervals, and the number of defectives recorded. The chart enables us to determine when the number or proportion of defectives is abnormally high and corrective action is demanded. It can also be used to detect when results are abnormally good, so that action or investigation can be instituted with a view to maintaining the improvement. Attribute charts must be designed to suit the process concerned and based on study of past performance.

In order to illustrate the principles involved we shall confine our attention to the case where defects are of a purely random nature. As an example, let us imagine that we are turning out small metal components which may be affected by blisters. Samples of 50 are taken from the production line at regular intervals and the number defective counted. Let us suppose that we find as a result of investigation that the average number of defectives in these samples is 2, at what stage should we take corrective action? According to statistical theory the incidence of random defects corresponds to what is known as the Poisson distribution. The figures in Table 5.4 show the probabilities of different numbers of defectives being found in our sample of 50, when the average or expected number is 2.

Table 5.4

No. of defects	Probability %
0	13·5
1	27·1
2	27·1
3	18·1
4	9·0
5	3·6
6	1·2
7	0·3

It will thus be seen that while anything from 0–4 defects will occur frequently in our samples and we shall occasionally get 5 or 6, a figure of 7 would be very rare unless something abnormal had gone wrong. It would therefore be reasonable to set a chart with a warning limit at 4 defectives and an action limit at 6, and to institute corrective procedure if the warning limit is exceeded in two successive samples, or the action limit in a single sample.

Figure 5.6 shows such a chart. The warning line is drawn between 4 and 5 and the action line between 6 and 7. Corrective action was initiated at point B, immediately a sample with 7 defectives was found. The corrective action took some time to become effective and the process returned to normal yield about point C. In this actual case, the machine did in fact go out of adjustment at point A, and as will be seen this was soon detected. The reader can verify for himself that the average

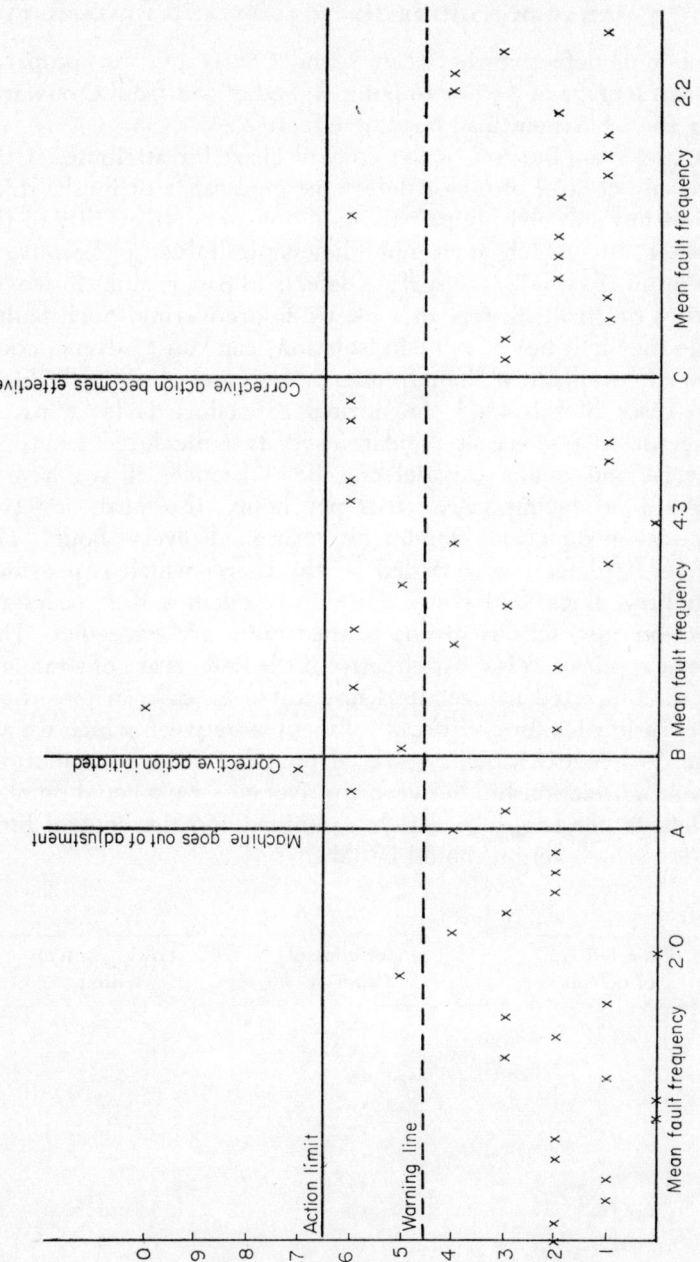

Figure 5.6. Attribute chart. Process out of control between A and C.

number of defectives between A and C was 4·3, as compared with an average of 2 prior to point A, and of 2·1 from C onwards after the adjustment had become effective.

A useful application of the control chart for attributes is the study and control of minor defects in appearance or finish which do not in themselves impair the function or serviceability of the product but which it is nonetheless desirable to keep to a minimum. Examples are surface defects in paper, slubs in woven textiles or small blisters in a plastic floorcovering. Such faults, while they may be tolerable in isolation, can cause adverse comment or complaint if their frequency becomes abnormally high.

In cases of this kind the normal procedure is for a patrol inspector to observe at regular intervals a measured length of material and count the defects. For instance, if we have a machine producing 3000 yards per hour, 100 yards, or two minutes' production, might be examined every hour. The number of defects is recorded on the chart, which is provided with limit lines, and corrective action taken and if necessary rejection instructions given, if the limits are exceeded. This type of application is most effective if the defects are of a random and disconnected nature. It is not suited to cases where there are occasional severe outbreaks of trouble between which we are clear of defects. Table 5.5, based on the Poisson distribution, shows the relationship between the average or expected number of defects per length or article examined and the control limit outside which action should be taken.

Table 5.5

Expected No. of defects	Upper control limit	Lower control limit
1	4	—
2	6	—
3	8	—
4	10	—
5	11	—
6	13	—
8	16	—
10	19	1
15	26	3
20	33	6

It will be noted that when the average number of defects is 10 or more there is a lower control limit. It might be thought that this is unnecessary as no-one minds making fewer defectives than normal, but if our objective is to improve performance, it is almost as important to investigate the cause of abnormally good results as abnormally bad. It may well be found that the institution of an attribute chart for defects and the investigation into causes which it encourages will result in a marked improvement, and limits can therefore be progressively reduced.

Control chart for percentage defective or percentage yield (p-chart)

We have seen that Shewhart control charts either for attributes or measured variables are a valuable tool of quality control where samples or measurements can be taken on the job and corrective action carried out quickly when the machine or process is getting out of control. In many cases, particularly in process industry, quality standards may be subject to longer-term influences. It is useful to be able to follow trends over a period of time and to chart results in such a manner that we can see clearly when significant progress has been made, or when a deterioration has occurred which requires investigation or corrective action. Yields of chemical processes may vary not only from batch to batch, but from week to week, and an adverse change if prolonged or progressive could be very costly. The percentage of scrap or waste produced on a production line will vary, and even if weekly figures are recorded and circulated to those responsible it is necessary to distinguish the important long-term trends, whether favourable or adverse, from the chance fluctuations from week to week. Again, if we are putting pressure on, or co-operating with, a vendor to improve the quality of his supplies, it may be useful to chart results in such a way that significant improvements can be clearly appreciated. It is in cases of this kind that the use of control charts for percentage defective or percentage yield can be of considerable value in helping those responsible to distinguish the significant from the incidental, and to direct efforts to trends which can have an important long-term effect on profitability, rather than waste time and effort on minor variances which may be self-correcting.

In order to illustrate the control chart for percentage defective we shall use the figures given in Table 5.6, which were recorded during 1966 in a factory, for the weekly percentage of scrap produced on a manufacturing process.

Table 5.6

Week No.	% scrap	Week No.	% scrap	Week No.	% scrap	Week No.	% scrap	Week No.	% scrap
1	15·9	11	14·1	21	10·9	31	10·7	41	8·7
2	14·4	12	12·9	22	8·7	32	7·8	42	9·7
3	17·9	13	15·5	23	10·8	33	6·3	43	9·2
4	15·4	14	11·4	24	10·1	34	13·3	44	9·0
5	20·8	15	9·8	25	12·8	35	7·0	45	7·8
6	17·7	16	12·2	26	10·7	36	8·8	46	12·0
7	17·3	17	7·8	27	12·5	37	10·3	47	8·2
8	20·2	18	8·2	28	11·2	38	13·2	48	10·9
9	16·7	19	10·3	29	11·2	39	11·8	49	9·5
10	11·4	20	8·6	30	13·0	40	11·0	50	8·6

The average percentage scrap obtained during the previous year had been 13·4. In order to distinguish random variability from the long-term trend, it is necessary to estimate the standard deviation of the results from the mean. In a case of this kind, where the mean is moving with time, there is a simple rule relating the standard deviation to the mean of the differences between successive pairs of results, namely:

$$\text{Standard deviation } (\sigma) = \frac{\text{Mean difference of successive pairs}}{1 \cdot 13}.$$

Obviously it would be necessary to make this estimate from previous records, and in fact this was done for the 1965 figures, which gave 2·3 for the mean difference between successive pairs and hence 2·0 for the standard deviation. The reader can easily verify for himself that the corresponding figures for 1966 are 2·4 and 2·1, showing that there has been little or no change in the inherent variability.

Figure 5.7 shows the figures plotted on a chart with warning lines, two standard deviations (4·0%) on either side of the standard. At the beginning of the year the standard taken was the previous year's average of 13·4 and warning lines were drawn at 17·4 and 9·4. Four readings above the upper warning line were obtained early in the year and action was taken to correct

CONTROL CHARTS

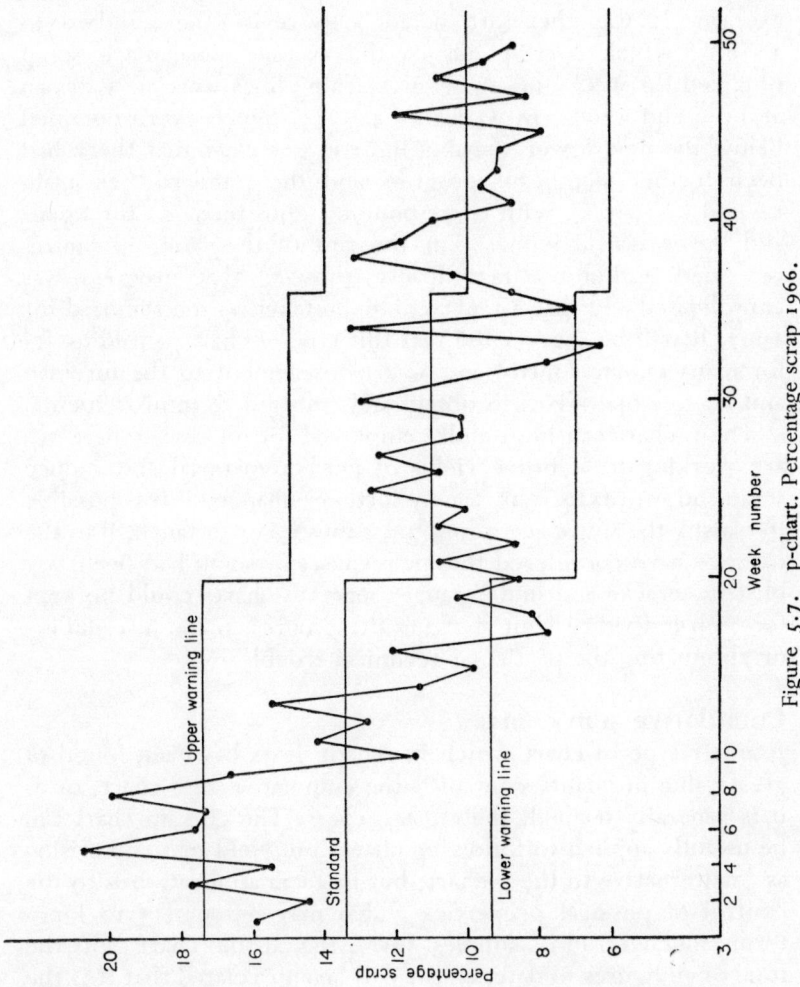

Figure 5.7. p-chart. Percentage scrap 1966.

the unsatisfactory position. In weeks 17 and 18, figures were obtained below the lower warning line and it was clear that the corrective action had been effective and that results were now being obtained significantly better than the previous year's average. It was therefore decided to revise the standard to 11·0%, which corresponds to the average percentage scrap obtained for weeks 11–20. New warning lines were now drawn at 15·0 and 9·0%. In weeks 33 and 35, figures were obtained below the new lower warning line; it was clear that there had been further significant progress and the standard was again revised to 10·0%, with corresponding adjustment to the upper and lower warning lines. For the rest of the year, all figures remained within the new limits, showing that progress was consolidated and that 10·0% could be taken as the standard for 1967. It will be appreciated that this type of chart proved useful for management control and an encouragement to the foreman and his key operatives to obtain and consolidate improvements.

The p-chart can be usefully employed for all cases where we are working to improve yields of perfect material and reduce scrap and waste. Its value can be further enhanced if it is possible to classify the waste according to its cause. For instance, if in the case we have considered the main causes of scrap had been say, blisters, cracks and under-gauge, separate charts could be kept for each individual fault showing the progress made in reducing or eliminating the particular technical trouble.

Cumulative sum charts

Another type of chart which in recent years has been found of great value in quality control is the cumulative sum chart, or as it is generally termed, the cu-sum chart. The cu-sum chart can be usefully applied to following changes of yield or scrap arising as an alternative to the p-chart, but its main application is to the control of physical properties which may be subject to long-term influences. In its simplest form, the cu-sum chart plots the total of all figures to date, and it will be appreciated that it is the gradient of the line which measures the value of the variable recorded. For instance, if we were recording the tensile strength of yarn, a series of high figures would give a line sloping more steeply than normal.

CONTROL CHARTS

The usefulness of a cu-sum chart is much enhanced if we plot not the cumulative sum of the readings themselves, but the cumulative sum of the differences from a target or standard figure, and pay special attention to the scale used. We shall then obtain a graph which is horizontal when results are on target, a positive slope when they are higher than normal and a negative slope when they are lower. Attention to scale will enable us to assess readily the degree of departure from target. We shall illustrate these features of the cu-sum chart by reconsidering the figures in Table 5.6. Our target or standard figure will be taken as the previous year's average of 13·4 and we shall plot the cumulative sum of the differences of the weekly figures from this target. The figures we require are shown in Table 5.7.

Table 5.7

Week No.	Weekly difference from target	Cu-sum of difference from target	Week	Weekly difference from target	Cu-sum of difference from target
1	+2·5	2·5	26	−2·7	−6·9
2	+1·0	3·5	27	−0·9	−7·8
3	+4·5	8·0	28	−2·2	−10·0
4	+2·0	10·4	29	−2·2	−12·2
5	+7·4	17·4	30	−0·4	−12·6
6	+4·3	21·7	31	−2·7	−15·3
7	+3·9	25·6	32	−5·6	−20·9
8	+6·8	32·4	33	−7·1	−28·0
9	+3·3	35·7	34	−0·1	−28·1
10	−2·0	33·7	35	−6·4	−34·5
11	+0·7	34·4	36	−4·6	−39·1
12	−1·5	32·9	37	−3·1	−42·2
13	+2·1	35·0	38	−0·2	−42·4
14	−2·0	33·0	39	−1·6	−44·0
15	−3·6	29·4	40	−2·4	−46·4
16	−1·2	28·2	41	−4·7	−51·1
17	−5·6	22·6	42	−3·7	−54·8
18	−5·2	17·4	43	−4·2	−59·0
19	−3·1	14·3	44	−4·4	−63·4
20	−4·8	9·5	45	−5·6	−69·0
21	−2·5	7·0	46	−1·4	−70·4
22	−4·7	2·3	47	−5·2	−75·6
23	−2·6	−0·3	48	−2·5	−78·1
24	−3·3	−3·6	49	−3·9	−82·0
25	−0·6	−4·2	50	−4·8	−86·8

In plotting our graph we shall adopt the following convention. The interval between weekly figures (2 mm) on the horizontal axis will be made equal to the equivalent of two standard deviations (4%) on the vertical axis. This will mean that, when results are two standard deviations above target, the graph will slope upwards at $45°$ and a downward slope of $45°$ will represent two standard deviations below. The results are illustrated in Figure 5.8.

Appreciating the fact that our percentage waste is measured by the gradient of the graph, it is easy to see that a marked improvement occurred at point A (week 9) and a further marked improvement at point B (week 13). At point C (week 24) there was a less marked falling off which was corrected at point D (week 31). It is easy to check whether such points of change in slope of the curve are significant, or whether they may be due to chance causes. This is very conveniently done by means of V-shaped masks which can be used to measure the change in gradient and determine its degree of significance. In fact it can be readily shown that the improvements at A and B are highly significant and that the improvement at D is significant as compared with the period between C and D, but not when compared with the period between B and C.

The following example illustrates the use of a cu-sum chart in the study of a physical property, which is subject to change with time. The actual study concerned the tearing strength of a heavy gauge paper, in which the purchaser had had trouble with a poor delivery and was demanding an improvement in standard. Samples were taken from each roll delivered and tested in the purchaser's laboratory. The supplier co-operated by providing information on the date of manufacture of each roll. The measurements of tearing strength are given in Table 5.8.

A study of past figures showed that the inherent variability of the product corresponded to a standard deviation of 3 lb, in an approximately normal distribution. The purchaser's requirements were that minimum strength of any paper received should be 10 lb. As we have seen, 95% of all readings in a normal distribution lie within two standard deviations of the mean, and virtually all within three. To ensure satisfactory supplies, therefore, a mean strength of 16 lb was essential and 19 lb

CONTROL CHARTS

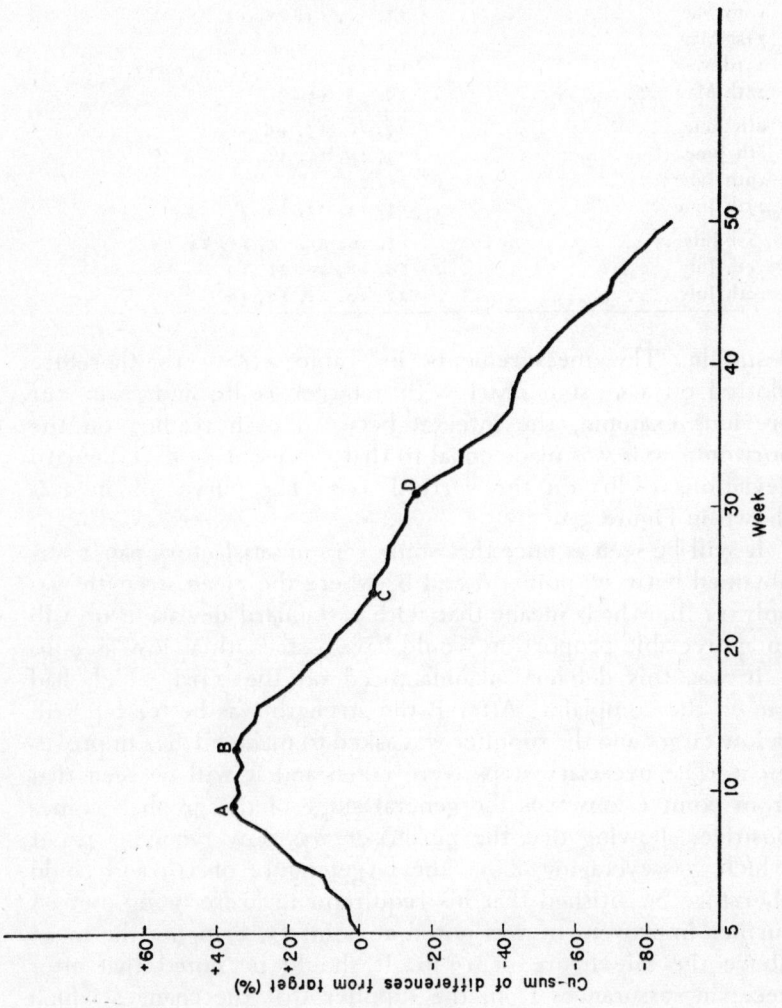

Figure 5.8. Cumulative sum chart. Percentage scrap 1966.

Table 5.8

Date	Tearing strengths (lb)
8th May	16
16th May	12, 11, 13, 11, 13, 16
21st May	19, 16
23rd May	13, 19, 10, 9, 15, 10, 8, 13, 12, 9, 10
28th May	15, 13, 15
6th June	17, 12, 13, 15
8th June	13, 16, 11, 14
19th June	19, 10
27th June	15, 14, 23, 14, 19, 15, 17, 26
1st July	13, 16, 21, 14, 15, 24, 20
5th July	26, 18, 16, 21, 19
10th July	22, 20, 21, 17, 19

desirable. The measurements in Table 5.8 were therefore plotted on a cu-sum chart with a target 16 lb, and as in our previous example, the interval between each reading on the horizontal axis was made equal to that representing two standard deviations (6 lb) on the vertical axis. The curve obtained is shown in Figure 5.9.

It will be seen at once that some very unsatisfactory paper was obtained between points A and B, where the mean strength was only 11 lb, which meant that with a standard deviation of 3 lb an appreciable proportion would have a strength as low as 5 lb.

It was this delivery manufactured on the 23rd which had caused the complaint. After B the strength was better but still below target and the supplier was asked to make further improvement. The necessary steps were taken and it will be seen that from point C onwards the general slope of the graph becomes positive, showing that the purchaser was now receiving paper which was averaging above the target figure of 16, and could therefore be satisfied that his requirements were being met. A further improvement was noted at point D, bringing the mean above the safe figure of 19 lb. It should be noted that after receiving assurances from the supplier that the changes which had given these improved results would be maintained, the purchaser was able with confidence to reduce the frequency of his testing substantially and, after a few months' satisfactory experience, to discontinue acceptance testing entirely apart

CONTROL CHARTS

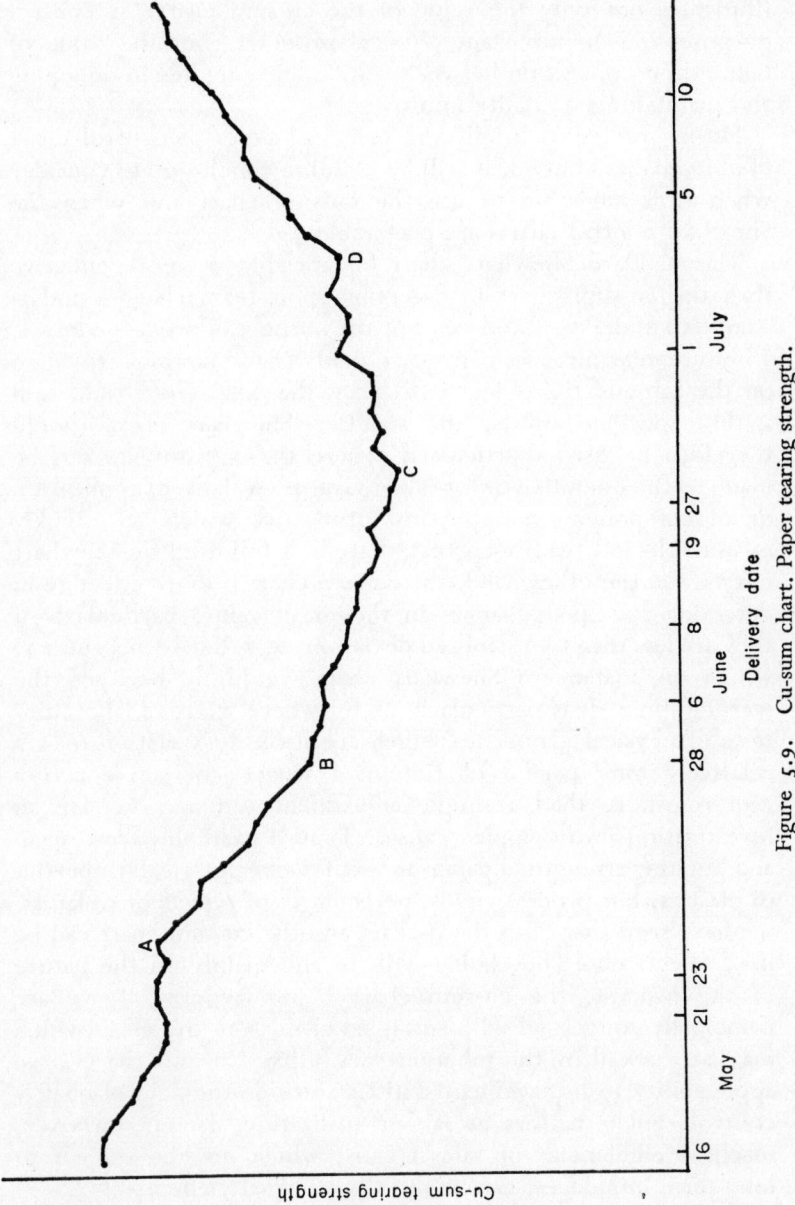

Figure 5.9. Cu-sum chart. Paper tearing strength.

from a weekly sample for record purposes. This case-study illustrates not only the value of the cu-sum chart for control purposes on an important physical property, but the value of technical co-operation between vendor and supplier in obtaining and maintaining a quality improvement.

Having reviewed briefly the principal types of control chart used in quality control, it will be useful in conclusion to consider when it is advisable to use the cu-sum chart and when the Shewhart control charts are preferable.

The standard Shewhart chart for variables is more effective than the cu-sum chart in detecting short-term changes and is simple to understand and use. For the purpose of process control, where regular measurements or quickly tested samples are taken on the job and the object is to keep the quality of production within specified limits, the standard Shewhart chart should therefore be used, particularly where the adjustments can be made by the operative or foreman. Gauges, weights of application or of components and physical properties which are quickly measurable and readily corrected are best followed on Shewhart charts. On the other hand, the cu-sum chart is more effective in detecting sustained changes in the mean value, particularly if they are less than two standard deviations, e.g. between $\frac{1}{2}$ and 2σ, which on a standard Shewhart chart would lie between the warning lines. It is particularly useful where we have laboratory tests of physical properties which are liable to variation over a relatively long period of time and where corrective action requires more than a simple adjustment and may involve an investigation into complex causes. Typical examples are tearing and bursting strength of paper or textiles and electrical properties of plastics. For process yields, percentages of rejects or of faults, we have seen that both the p-chart and the cu-sum chart can be used effectively. The choice will depend mainly on the nature of the process, the cu-sum chart being favoured if we are principally concerned with sustained changes in the mean which may be masked by the inherent variability. Cu-sum charts have applicability to management statistics outside the field of quality control. Such matters as labour utilization, labour turnover, machine efficiencies or sales trends, which may be subject to long-term influences, can be dealt with equally effectively.

6 Design for Quality and Reliability

'A good reputation for well designed goods or components, fit for the purpose, which don't fail or break down, is the criterion for certain success. A bad reputation is a very costly luxury which this nation cannot afford.'

H.R.H. Prince Philip, Duke of Edinburgh

IN OUR INTRODUCTORY CHAPTER we summarized an eight-point plan for a logical quality and reliability system. The first four points, which we shall now discuss in more detail, were:

(1) Study of customer requirements with due consideration to performance and price.
(2) Satisfactory design of product, or service, thoroughly proved by testing in order to establish its reliability under the conditions to which it will be subjected in use.
(3) Full specification of the requirements of the design, which must be clearly understood by everyone concerned with the production of the constituent parts and of the complete end product.
(4) Confirmation that the manufacturing processes are capable of meeting the design requirements.

The quality and performance of a product depends on two main factors:

(a) Correct designing for the purpose intended and the conditions to which the product will be subjected.
(b) The standard of conformity with the approved design.

Failure in the field and consequent bad reputation of the product may result either from deficiencies in design which give a product which simply does not fulfil its function, or from poor conformance with a design which in itself may be satisfactory. Too much emphasis is often placed when considering quality control on the inspection function. Quality cannot be inspected into a product, it must first be designed and then manufactured into it. It is essential to good design that not only will the product fulfil its function when correctly made, but that a high standard of conformity without unnecessary costs and high rejections can be obtained.

At the pre-design stage it is important that as full information as possible is available about customers' requirements, with due regard to performance and price. It is important to study standards which may have already been set by competitors and also, if the new design is to be a replacement of one of the company's existing products, we must appreciate clearly what improvements or economies are desirable. As regards information from the market and study of customers' requirements, the problems are very different if we are dealing with capital equipment or manufactured intermediates supplied to other manufacturing companies from those which arise in the case of consumer goods, where the ultimate customer is the general public. A third case with its own special problems is that of contracts in consumer durables where the customer is a government department or large corporation. The first case is in some ways the easiest to deal with. The desirable feature here is to cultivate the partner relationship between the supplier and the customer and to provide facilities for direct contact on technical matters between those in each company who can directly influence the course of action. Where the ultimate customer is the general public, we are usually faced with the position that supply is not direct, but through one or more intermediates in the distribution chain. The feed-back of information may be blocked or distorted, particularly if it has to pass through non-technical channels. In the third case, particularly where government departments are involved, conformity with a detailed specification may be a prime requirement, and while this has been normally drawn up as a safeguard of quality, it may well become a brake on progress and improvement if the necessary direct technical contacts are not made and used. Finally, we have the special problems of the export trade, where requirements and user conditions which affect our design may be very different from those met with on the home market.

It cannot be too strongly emphasized that it is the combination of performance and price which is important. The market for the high-priced prestige article, designed for quality and reliability alone, is today very limited. On the other hand, the cheap and unreliable product, however strongly promoted, will soon be ousted by a competing article which is more dependable. Successful sales will, in the long run, go to the product which gives the best value for money. It is useful to think of the ratio quality and reliability to product cost and to appreciate that the object of maximum value for money can be achieved not only by improving product quality, but by reducing cost by elimination of those items of expenditure which contribute nothing to the function, performance, reliability and sales appeal of the article.

Having satisfied ourselves that we know what the customer requires, the price range in the market at which we are to aim, and the qualities and performance the customer will expect in return for his money in that price range, we can now proceed with design, bearing in mind three cardinal requirements which will be essential to success and profitability.

(i) Inherent weaknesses which can result in premature failure in the field must be avoided.

(ii) Design must be co-ordinated with production engineering and manufacture to minimize the practical difficulties of producing consistently what is required, with the minimum wastage and scrap.

(iii) Design must avoid any unnecessary cost which contributes nothing to the saleability and serviceability of the product. In this connection careful attention must be paid to the choice of materials, often the major item in the total variable cost, and to the use of new methods and modern techniques where these are justifiable.

It will be appreciated from the above that except with a very simple product or in a small organization it will be quite impossible for any one person to possess today all the technical and specialized knowledge to achieve such requirements in isolation. Hence we require a design team, not a team of designers but of designer or designers co-ordinated with other skills, particularly those of production or methods engineering, quality control and purchasing.

Let us now go on to consider in a little more detail these cardinal requirements and how they can be achieved.

Avoiding inherent weaknesses and premature failure

Whatever the price range at which we are aiming, the customer will expect a commensurable period of reliable service. Garments which shrink badly on the first laundering or curtains which fade during the first sunny spell may sell at a low price for strictly temporary use, but no-one will repurchase or recommend them to a friend for normal service. A good example of a well designed, simple and reliable product is the Camping Gaz International, a small stove used by campers and picnickers. The plastic stand is of a quality which stands up to repeated assembly and dismantling, and all other parts are of a similar standard. The main assembly and burner will give many years' use without appreciable maintenance, and the gas containers are uniform and a reliable fit. As a result a flourishing export trade has been built up and purchases can be made in most parts of the world. The tourist, whether he spends his holiday in Western Europe, Ireland, England, Canada or the western States of America, can take his burner with him in the confidence that it will work, and that if he runs out of gas containers he will be able to purchase them on his travels.

In the case of a much more complex assembly such as a car, complete freedom from trouble in the initial stages is difficult to ensure and some allowance may be made for minor failings, which can be put right on the first servicing, but defects which cause danger or serious inconvenience must be avoided. The Rover Company of Solihull classifies the seriousness of defects into four categories:

(i) Faults involving safety of the vehicle and occupants
(ii) Faults which would affect company prestige

(iii) Faults which are costly to correct
(iv) Minor faults which can cause customer irritation.

The higher the priority of the defect, the more thorough must be the tests and precautions to ensure reliability. Where early failure of one component could endanger life or limb, or seriously affect the company's reputation, the most stringent precautions are taken to minimize the risk. Category (iv) defects are, however, by no means ignored. The initial guarantee and good liaison with agents ensures that information fed back to headquarters is as complete as possible, and numbers and percentages of complaints are charted as shown in Figure 6.1.[6] The most frequent complaints, even if of a minor nature, can then receive priority and design weaknesses be corrected.

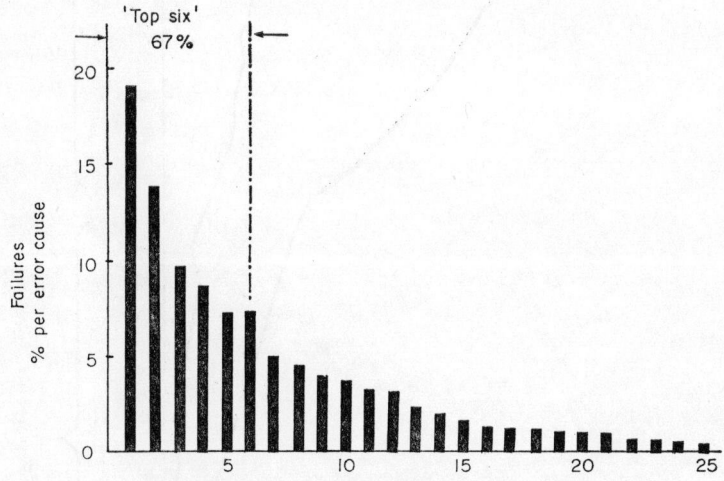

Figure 6.1. Different error causes. Reproduced by permission of the Rover Company, Solihull.

Figure 6.2[7] illustrates the effect of such feed-back of information, showing the great improvement obtained in twelve months with regard to a certain defect which had affected over 1% of the 1964 cars during their first year and 5% by the end of the second. As a result of action taken, the 1965 production

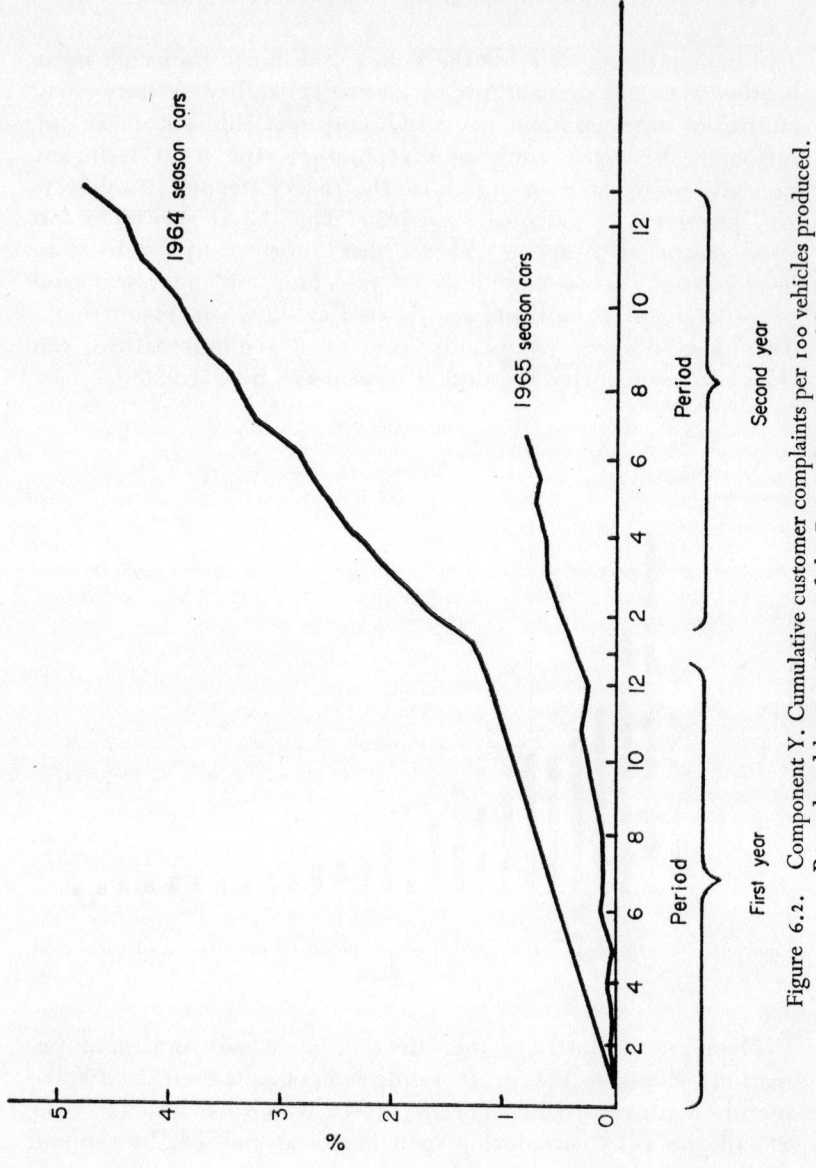

Figure 6.2. Component Y. Cumulative customer complaints per 100 vehicles produced. Reproduced by permission of the Rover Company, Solihull.

was almost clear of the trouble during its first twelve months, and defects had not reached 1% midway through the second year of the cars' life.

Rover has a comprehensive quality and reliability control organization responsible to an executive director and comprising a number of specialist sections which include a materials laboratory, quality engineering, bodies and fitments, inspection, reliability and overseas quality. The inspection and test procedure includes a run for all vehicles at least twice round a four-mile track, after which any car which does not behave satisfactorily or gives indication of possible trouble is returned to the shops for adjustment or correction.

It is important to evaluate the reliability of a product as far as possible while still at the design and pre-production stage and minimize the risk of waiting until it is in the customer's hands and so risking the company's reputation. To this end production samples should be tested under simulated service conditions. Simple statistical methods are very useful in reliability testing of this kind as can be illustrated by the following example for which I am indebted to Mr. M. T. Witts of the Rover Company.[8]

Two electric switches designed for use in cars were tested for reliability by taking random samples from the production line over a period and testing each switch to failure. The initial test was carried out on ten switches of each type. Even these small samples enable one to make a reasonable estimate of the mean life to be expected from the switch and the degree of variability as measured by the standard deviation from the mean. The distributions of life are shown in Figure 6.3. In type A the mean is more than three standard deviations above the design life, and we can be reasonably satisfied that in practice there will be little risk of premature failure. In type B, on the other hand, the spread is greater, the margin between the mean life and the design life is only about one and one-third standard deviations, and there is clearly a risk that we may have about 10% of premature failures. A further test was therefore made on type B with 50 switches, which confirmed the dangers indicated by the first sample. It was therefore clear that while switch A could be accepted as reliable, a modification to switch B was necessary and the design was carefully re-examined to eliminate the weakness.

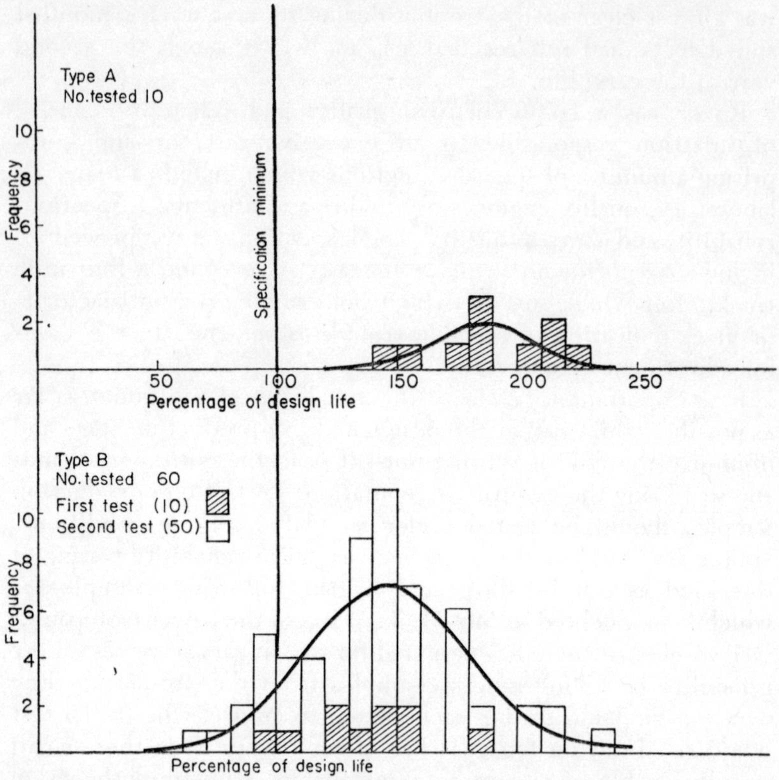

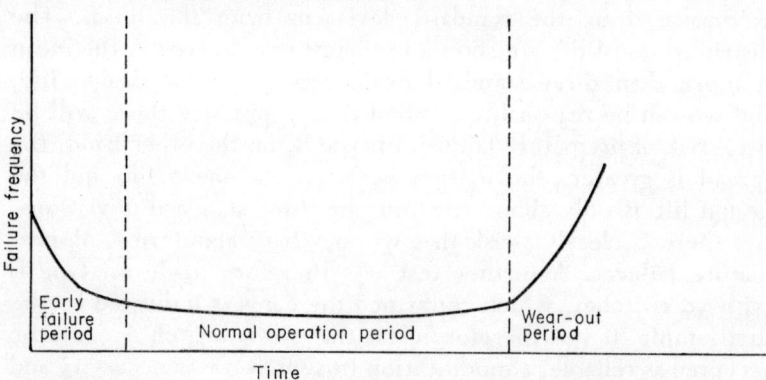

Figure 6.3. Life distribution of electric switches.

Figure 6.4. Another type of failure frequency distribution.

DESIGN FOR QUALITY AND RELIABILITY

In the above example the reliability tests gave a simple distribution, not far from normal. In other cases the distribution may be of a different type. One pattern frequently encountered in reliability tests, for example in the electronics industry, shows three distinct phases, a high rate of early failure which drops off rapidly, followed by a fairly constant failure rate during the product's normal operating period, and finally a wear-out period in which failure rate again rises rapidly, as illustrated in Figure 6.4.

Co-ordinating design with production engineering and manufacture

This is an extremely important consideration if we are to be successful in producing a high proportion of satisfactory output and avoiding high costs of rejection scrap and rework. Is the plant and are the operatives capable of producing consistently the requirements of the design? If not, can either design or manufacturing technique be altered to overcome the difficulty, or are the originally specified tolerances necessary to efficient function in the assembly? Process capability studies should be made to ensure that one's own plant is able to turn out what is required with consistency, and suppliers must also be assessed to determine their capability with regard to consistent supply of satisfactory parts. These investigations should be carried out during the pre-production stage, when the cost of making modifications will be far less than when large quantities of material are going through the production lines. The procedure clearly involves close liaison between the designer, production engineer, purchaser and the quality control specialist.

Simple statistical methods such as we discussed in Chapter 4 can be very useful in process capability and design improvement studies. If a number of sample parts are made under carefully supervised conditions and a distribution of measurements obtained in which the spread of results is wider than the specification limits, then we know that a proportion of defective material will be inevitable when the part is put into production. Similarly, if a component is giving rise to occasional complaints, a statistical study of sample components from the production line will show where design or process modification is needed.

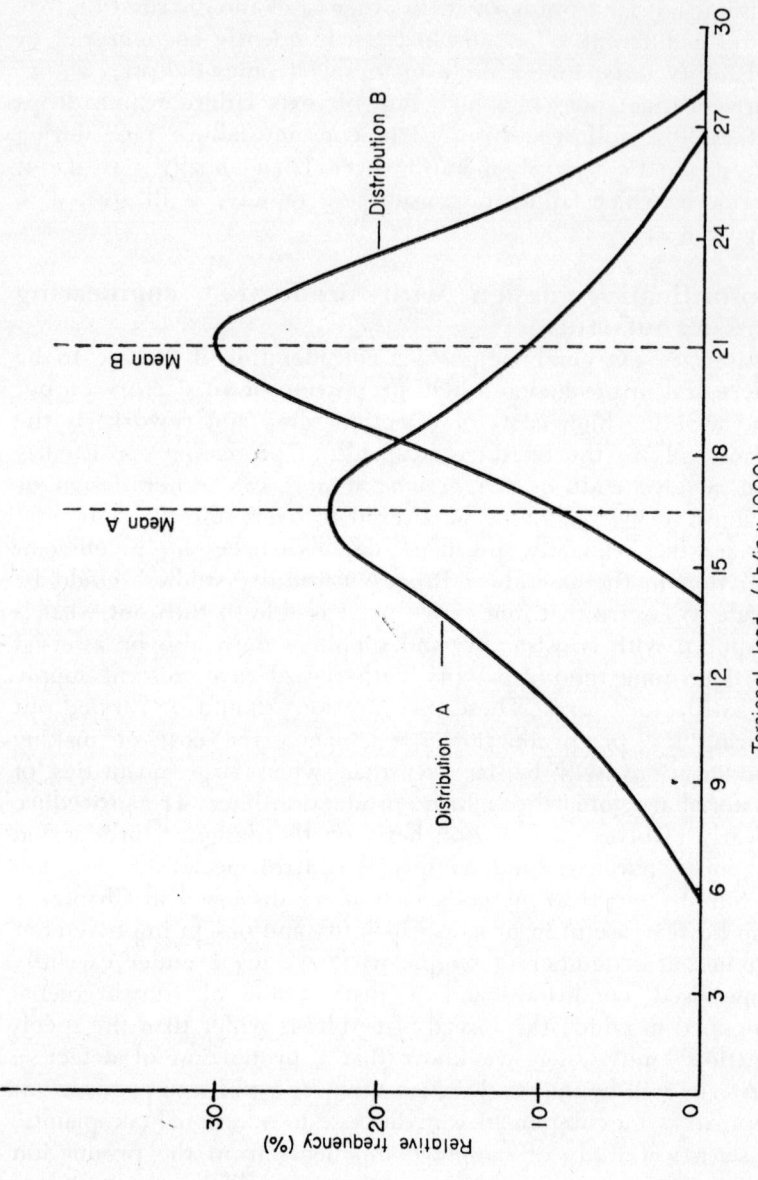

Figure 6.5. Component Y. Strength variation. Reproduced by permission of the Rover Company, Solihull.

Figure 6.5 illustrates this for a certain component which, while not critical, was causing some trouble with cracking after accidental impact. A number of regular samples were therefore withdrawn from production for examination and test. The load at which each unit failed was plotted over a period and the distribution curve A obtained. As will be seen from the diagram, there was considerable variation in strength. As a result a change to the heat treatment specification was made and sampling of production components continued. The failure load was again plotted over a period and distribution B was obtained. This showed an improved average strength and, what was equally important, the spread of results was considerably reduced. It could now be assumed that the production plant was able to produce consistently a reliable article.

When capability studies are undertaken it is often found that limits and tolerances have been set unnecessarily tightly for the job the component has to perform, but in other cases redesign or modification to plant is shown to be necessary. The institution of such studies is essential if we are to embark on full-scale production without the prospect of producing substantial quantities of defective output. The advantages of the statistical approach will be very clear; it is not only the mean value which is important. To avoid the danger, or in fact the certainty, of producing considerable quantities of defective work with all the expenses involved in inspection, rework and scrap, we must make sure that the plant to be used is capable of producing to the design requirements allowing for the spread of results which will occur on either side of the mean figure.

Avoidance of unnecessary costs which contribute nothing to the serviceability and saleability of the product

Shortly after the last war the term *value analysis* was coined by the General Electric Company of America to describe the set of techniques, created by that company, for continuous reduction of manufacturing costs, producing at the same time goods of equal or if possible higher quality. Lawrence D. Miles was given the responsibility for their value analysis programme and is the author of the well-known textbook *Techniques of Value Analysis and Engineering*, which will give the reader a very clear picture of the principles and techniques involved. Some authorities draw a distinction between the terms *value analysis* and *value engineering*, applying the former to the study of an existing product and the latter when we are concerned with the original design. In practice it is difficult to draw a clear distinction, as the analysis and study of existing product may result in considerable redesign, and new designs are more often than not adaptations or replacements of ones used for earlier or competitive products. The terms, therefore, are frequently used synonymously.

It is not only in the engineering industry that these principles and techniques can and should be applied. It is just as important to design the formulation of a chemical product or a range of printed patterns with a view to giving the customer the best value for the money he spends and the manufacturer the maximum margin of profit, as it is in the case of an engineering assembly such as a car or a television set. For want of a better term, therefore, we shall use *value analysis* to cover the techniques whether applied to an existing, redesigned or new product.

Value analysis has been defined as a technique for co-ordinating the skills of design, purchasing and production to minimize all controllable elements in the cost of a product, subject to the condition that it will fulfil its function adequately from the point of view of the customer. In this book, where our prime consideration is quality and reliability, let us recast this definition to alter the emphasis and describe value analysis as a technique for co-ordinating the skills of design, purchasing and production to ensure that a product, whether new or existing, will fulfil its function adequately and reliably from the point of view of the customer, subject to the condition that all controllable elements of cost which do not contribute to that objective are minimized. With this definition in mind we shall review the essentials of the technique and illustrate it with case studies, where the outcome has not only been cost economy but better value to the customer.

In value analysis a fundamental factor is team work, the bringing together of different skills, particularly those of design, purchasing, production and cost analysis, to bear on the problem under consideration. It is important that the members of the team should be departmental executives, the only full-time specialist being the team chairman. Regular meetings are arranged and the chairman is responsible for seeing that the data required for each meeting are carefully prepared, and for directing and progressing the work. The ideal size of team normally recommended is four or five, and for engineering assemblies, design engineer, production engineer and purchasing officer are essential participants. Other skills such as costing, quality control and marketing will be brought in when required, but for efficient work it is important not to let the regular team become too big. The method of working will of course depend to some extent on the nature of the product or design being studied, but the following scheme may be regarded as typical.

For the first meeting of the team, data must be carefully prepared and will depend on whether we are concerned with the design of a new product, or with the redesign or modification of an existing product. For an existing product we should require:

(1) The assembled product.
(2) Set of components.
(3) Samples of raw materials.
(4) Scrap layout.
(5) Competitor's products.
(6) Details based on market experience of any weakness, or common cause of complaint, or premature failure.
(7) An evaluation sheet giving the following information:
Material specification
Raw material supplies
Order quantities
Manufacturing operations
Labour costs
Raw material and finished weights and unit costs
Details of material waste with costs

For a new product the initial analysis will be similar, except that drawings, trial material or models would have to take the place of the assembled product and its components. Item (6) becomes:

> Details based on market experience or reports of similar products whether our own or competitive; foreseeable dangers or weakness to be guarded against.

Each component and aspect of the manufacturing process is then tackled with the aid of a systematic questioning sequence on the following lines.

(1) *Subject—Function*
   Basic question: What functions are performed by the component?
   Analysis: Are all functions essential to the assembly? What other ways are there of achieving the same function? Can all or any of the functions performed be incorporated in another component?

(2) *Subject—Material specification or Material content*
   Basic question: What is the full material specification?
   Analysis: Can any dimension be reduced? Is the part over size? Can a less costly material be used?

(3) *Subject—Material waste*
   Basic question: What percentage of material is wasted?
   Analysis: Can waste be reduced by making the blank nearer to the finished size? Can waste be reduced by minor modifications? Can waste be reduced by changing the method of manufacture?

(4) *Subject—Standardization*
   Basic question: To what degree are we using standard methods and components?
   Analysis: Can our range of manufacture be simplified without loss of business? Is the part or component of standard manufacture?

If not, can it be replaced by a standard part or component?

Can any part or component be rationalized with one used in another product manufactured by the company?

(5) *Subject—Limits and tolerances*
Basic question: What are the existing limits and tolerances?
Analysis: What tolerances are necessary for the proper fitting and functioning of the component?

Are these tolerances compatible with existing production methods, machine tools and plant?

If not, can methods or machines be adapted to comply with the necessary tolerances?

It is not uncommon practice for designers to play safe and specify tolerances which are unnecessarily narrow. This practice is soon appreciated by production and inspection supervision and material marginally outside the specified limits is permitted to be produced and passed. It is far better to specify limits and tolerances realistically and then insist on strict conformity with specification. If the dimensions of a component are not critical recognize the fact, and much more close attention will be paid to those which are really important and require precision.

A similar questioning sequence is applied to other aspects of the design, such as surface finish, direct labour costs and direct material costs.

Wherever our original analysis shows that there is a danger of weakness, complaint or premature failure, an additional question must be asked. Can the function or performance be improved by suitable modification?

At this speculative stage of the investigation, suggestions must be invited and encouraged from all quarters and it is an important rule that every member of the team is free to make suggestions, whether on his own speciality or not. As a result of discussion and examination of costs, the ideas which offer the most promise are then selected for further examination. The remaining stages of the study are development, recommendation, implementation and testing. Most of this work takes place outside the formal meeting and will probably take much longer than the initial analysis. Hence, it becomes possible to bring a series of components or processes under analysis and speculation while others are being developed, implemented and proved. It may well be found in the course of the investigation that substantial savings can be made which may permit the use of a better and more costly material to remove a weakness or give improved service and at the same time make an overall economy.

In drawing up the process specification, with particular reference to labour utilization and process timing, it is useful to use the method study technique and prepare an operational process chart while the product is still at the design stage. Method study on the drawing board is far more economical than having to make costly changes or alterations to layout at a later stage. It is here that we should be able to satisfy ourselves that design is suited to good quality control and whether and where inspection operations are necessary and economical.

When value analysis has been firmly established in a company, it becomes a way of thought rather than a specific technique and spreads throughout the organization. Many good suggestions and ideas originate and are developed outside the formal meetings. It remains important, however, to maintain the regular meetings to progress and co-ordinate the work.

One problem in design for quality and reliability combined with economy is the time factor. The decision as to when to commit design to production is crucial. A product launched too quickly may well be a disaster technically, but if too long delayed it may be a failure commercially. Very often there are strong pressures once an idea has been formulated, or a change asked for, to minimize the time from the drawing board to full production. It may be argued that the principles and techniques we have been discussing are all right in theory, but in practice will take too long. A great deal can be done by thinking ahead and anticipating future requirements, and by systematic planning once the decision to go ahead has been taken. The alternative is so often an inferior or troublesome product, failure in any case to keep to the promised dates and unnecessary costs caused by excessive rejects and complaints. Some of the most successful design achievements have not been unduly hurried. Work was started on the Rover 2000 in 1957, nearly seven years before the car was put on the market.

Published work on value analysis is very largely confined to the engineering industry. The reader will find many excellent examples clearly described in Miles's book. The following two case studies are given to show how the same principles have been applied in other industries.

(1) Design of envelope
The envelope is the ordinary 'Banker' type $3\frac{1}{2}$ in. × 6 in. The original design was as shown in Figure 6.6, which illustrates the unfolded shape and the finished envelope. The net area of paper required was 49·0 sq. in., but it is impossible to cut the shapes without considerable waste. The shapes were die-cut from stacks of paper in sheets of appropriate size, eight shapes from each sheet, and Figure 6.7 shows the most economical method which the production engineer could devise to suit the design. Allowing for the fact that to obtain a clean cut it is necessary to allow about $\frac{3}{16}$ in. between each cut and between the edges of the die and the outer edges of the paper, the minimum percentage waste was approximately 16%. The gross area of paper required per envelope was thus 57 sq. in. Die-cutting of the sheets, and the folding and gumming of the envelope, were by necessity separate operations.

Figure 6.8 shows the new design, in which designer and production engineer co-operated to make it possible to produce an envelope direct from a reel of paper with a minimum of cutting waste. The basic shape is a rhombus, and Figure 6.9 shows that the only waste now involved is the small pieces marked A, B and C removed to permit folding and the necessary overlap. The shapes can therefore be cut directly from reels of paper $6\frac{1}{2}$ in. wide, and as die-cutting is no longer involved, it is now unnecessary to leave any gap between successive cuts. Furthermore, it is now possible to carry out the whole process, from reel to finished envelope, in a single machine operated by one girl. The gross area of paper per envelope was reduced from 57·0 sq. in. to 49·5 sq. in., and the minimum waste reduced from 8 sq. in. or 16% to 0·65 sq. in. or 1·3%. After allowing

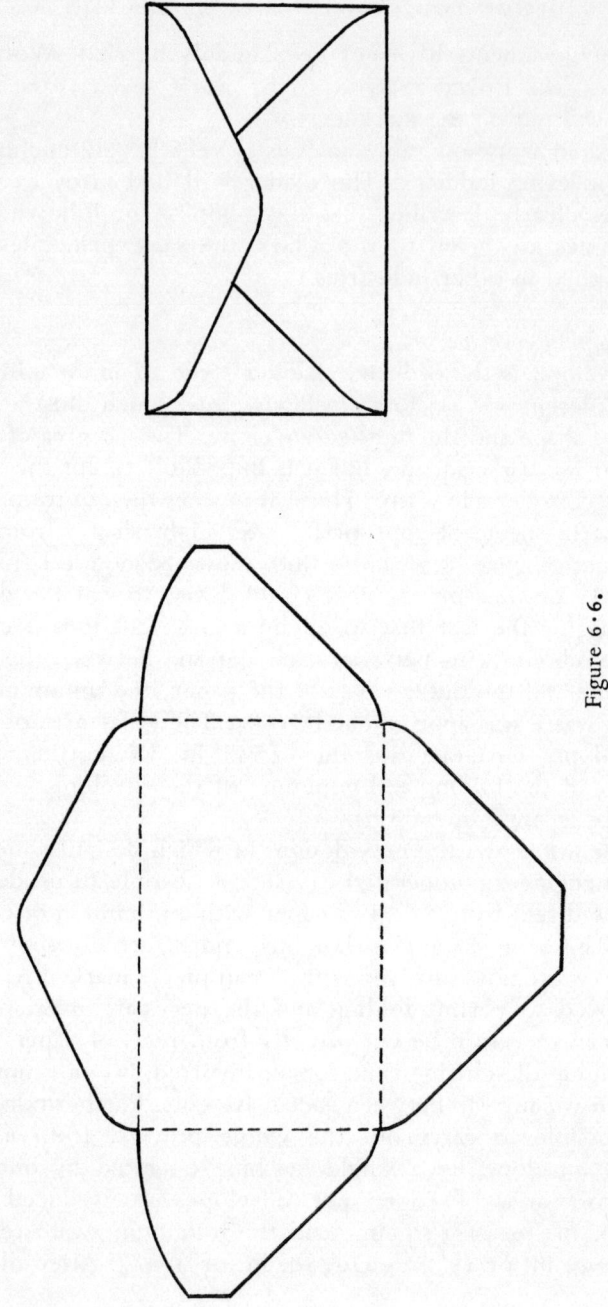

Figure 6·6.

DESIGN FOR QUALITY AND RELIABILITY 93

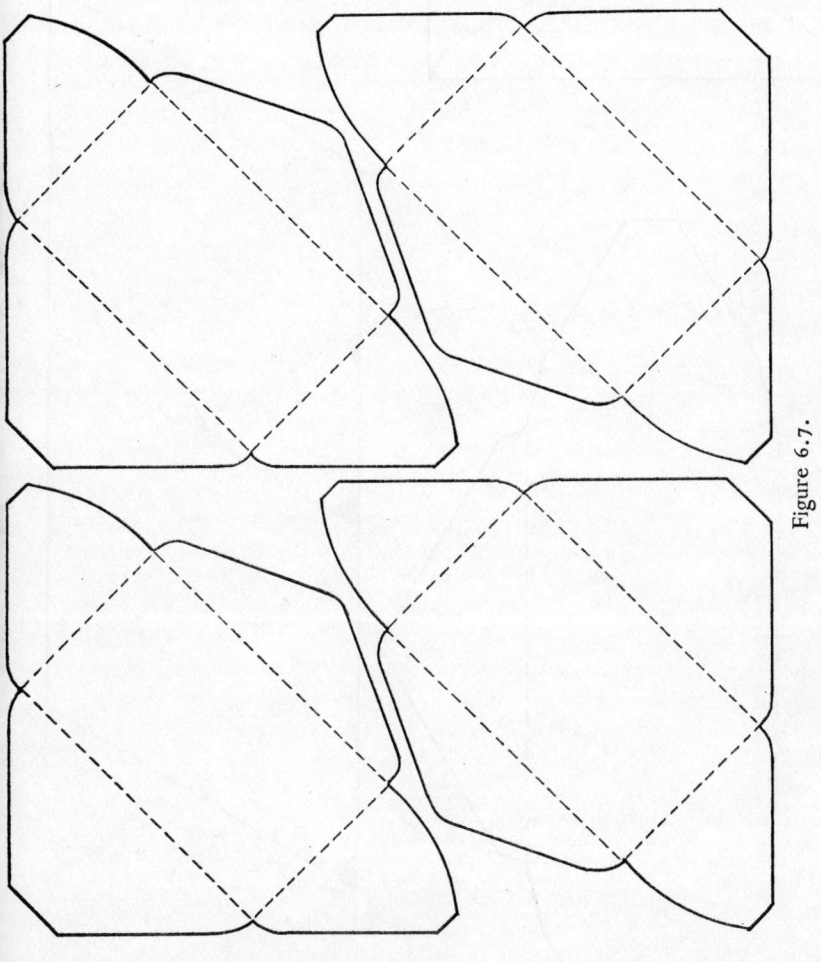

Figure 6.7.

Figures 6.6 and 6.7. Envelope—original design.

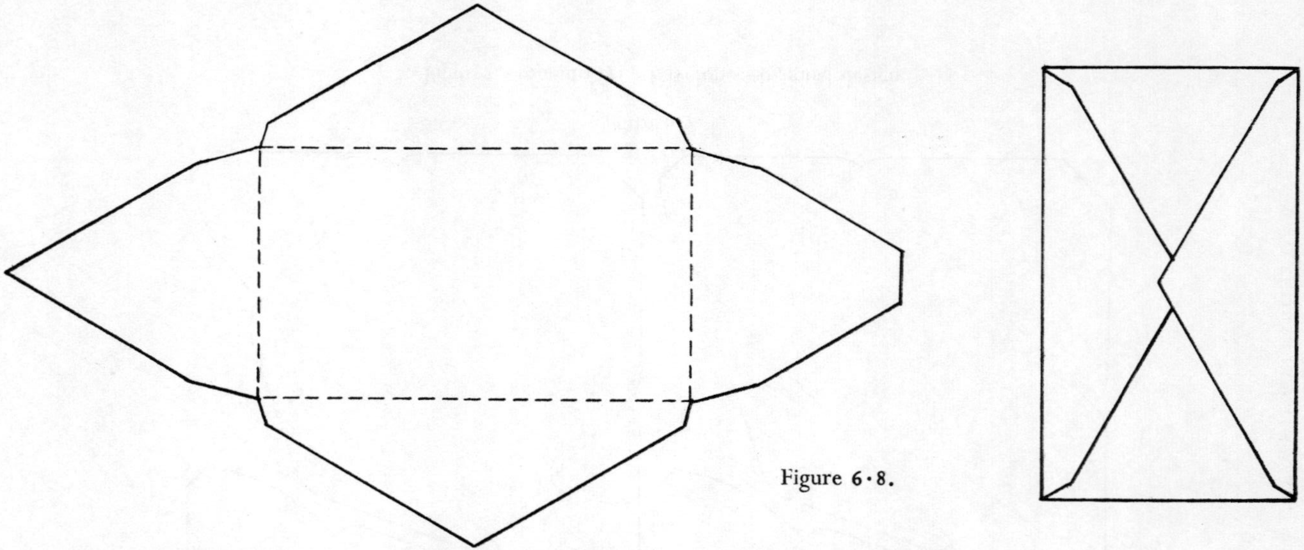

Figure 6·8.

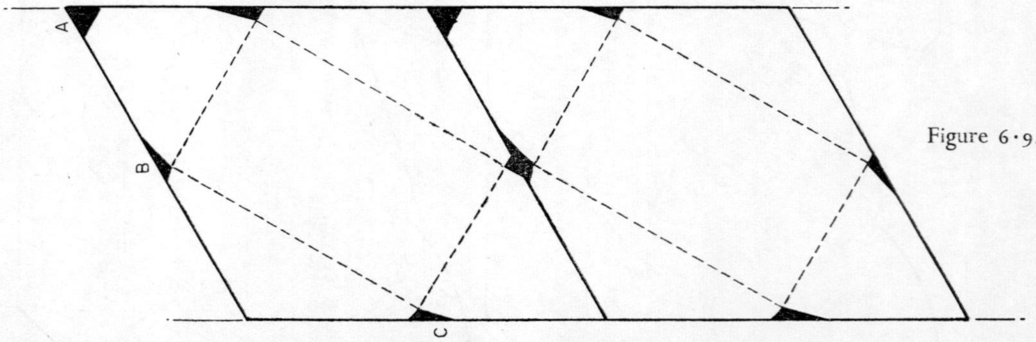

Figure 6·9.

A B C

Figures 6·8 and 6·9. Envelope, new design.

for the salvage value of the waste paper, roughly 25% of its delivered price, and a small saving in the cost of paper through purchasing in reels instead of sheets, the redesign thus made it possible to produce an envelope of quality equal to the original with a saving of over 10% in material cost, together with appreciable saving in labour.

(2) *Packing of floorcoverings*
This case study relates to the packing of a vinyl floorcovering, and resulted not only in a useful cost saving, but in improved protection for the product and a more attractive pack.

The original packing consisted of a chipboard carton, 8 in. in diameter and 48 in. in length, with slip-on chipboard lids with 2 in. rims. The cartons contained a roll of floorcovering, the total weight being 125 lb. The function of the lids was to close the carton and give protection to the product during transit, and also to provide a firm base on which palletized loads, up to three high, could be stacked. The cost of the chipboard lids was $8\frac{1}{2}$d. There could be no question of reducing quality, as with the existing lids a certain amount of damage occurred, particularly in cold weather, and was almost invariably confined to the ends of the roll. The damage was clearly caused by careless unloading from transport vehicles, but, in spite of warning labels and representations to the transport contractors, it could not be eliminated.

During the 'speculation' stage of the investigation, plastic (polythene) was proposed as an alternative material, and it was suggested that if the bases were recessed, the extended rim keeping the floorcovering clear of the floor and absorbing the shock when pieces were dropped, risk of damage would be eliminated.

After further discussion and trial, a plastic lid was designed in which the base was recessed $\frac{3}{8}$ in. and the depth of the rim was reduced from $2\frac{1}{2}$ in. to $1\frac{3}{4}$ in., which was found adequate for securing the lid to the carton and reduced the quantity of material required. The recessed lids stood firmly on the floor and were entirely satisfactory in palletization. Trials involving standing the rolls overnight at a low temperature and dropping from the height of a transport vehicle on to a concrete roadway

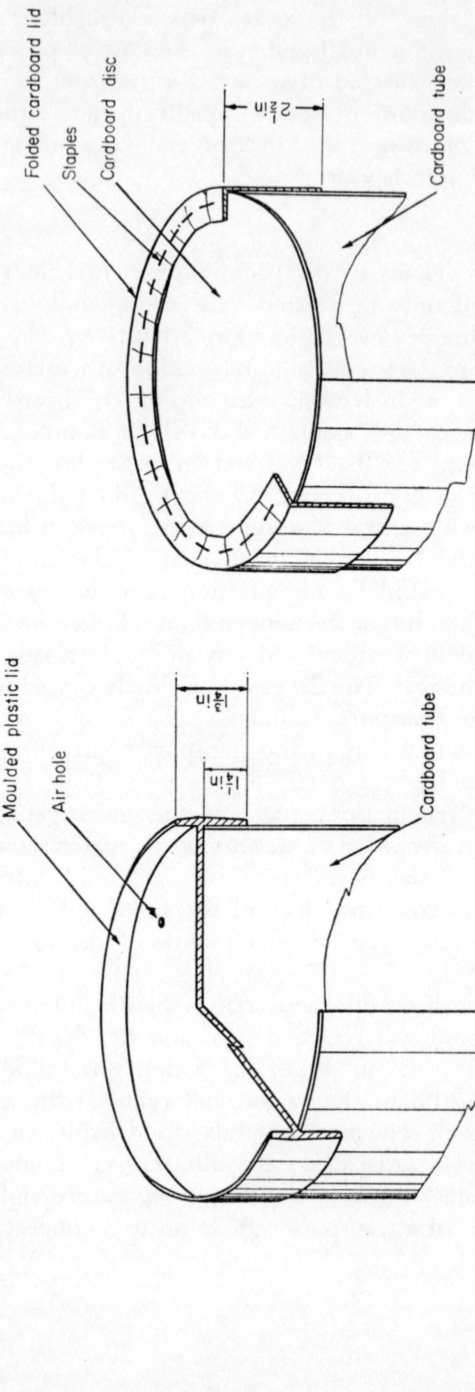

Figure 6.10. Redesign of carton lids.

showed that end protection had been outstandingly improved. However, the cost of the new lid was 9½d., or 1d. more than the original.

The purchasing department arranged for this lid to be supplied and the change was successful, transit damage complaints through the ensuing winter being virtually nil. The matter of cost was closely followed up by the buyer with the supplier, who suggested in the course of discussion that if black lids were acceptable, it would enable him to use a high proportion of scrap and the price could be considerably reduced. Samples of the black lids were supplied for trial purposes and proved equally as effective as the white translucent lids originally supplied. The cost of the black lids was 1¾d. less than the white plastic lids and ¾d. less than the original chipboard.

Several hundred thousand lids were required annually and the combined savings through cheaper lids and the elimination of transport damage claims came to approximately £2000 per annum. The change is illustrated in Figure 6.10.

Value analysis, far from being a tool for 'cheapening' in the bad sense, is a most powerful means of improving quality, particularly when we consider the ratio of quality to cost. Producing an article or component equal in performance and appearance at lower cost than the original is improving this ratio and if, as is frequently the case in these competitive days, part of the saving is handed on to the customer, gives him better value for the money he spends. Very often, however, the result is an actual improvement in quality and performance. Simplification of an assembly and a reduction in the number of components will frequently result not only in a saving of production cost, but in less danger of failure and a product of greater reliability.

It is no coincidence that the General Electric Company of America and Joseph Lucas in Great Britain are leading exponents of both value analysis and total quality control. When we are concerned with design and value for money and reliability of the product, the techniques are both complementary and closely interlinked.

7 Organizing for Quality and Reliability

'If you can measure that of which you speak, and can express it by a number, you know something of your subject, but if you cannot measure it, your knowledge is meagre and unsatisfactory.'

Lord Kelvin

IT IS DIFFICULT to imagine any firm that is still in business which has no system for control of quality, so in discussing organization our problem is mainly one of improvement rather than introduction. On the other hand, the existing system may well be old-fashioned, costly or inefficient, and radical changes and a new outlook may be required. It is not uncommon for the whole emphasis for ensuring product quality to be centred on the inspection function, whereas the battle for quality and reliability at the minimum cost should pervade all departments of the business. Again, there may be ignorance and consequent suspicion of modern techniques and methods, which if they were employed could be much more effective and less costly than existing procedures.

In no two companies are circumstances alike, and while it is impossible to lay down rigid rules as to how we should organize for improved quality control, certain general principles are well worth discussing and can help us to move towards the desired objectives. Modern research is leading to the conclusion that there is no standard pattern for organization of the successful business, and that success depends far more on personal relationships, team work and morale than on the details of organizational structure. Successful results, even in similar types of industry, are obtained with widely differing organizational schemes. This in no way diminishes the importance of organization, but recognizes the fact that human personalities and abilities vary so widely that what works best in one situation may differ considerably from what will give the best results elsewhere. What is true of business organization in general applies to organization for quality.

One thing is indisputable: control of quality to achieve maximum success must embrace the business as a whole, and the attitude of management at the top must be right. Top management must not only be interested and be seen to show interest, but needs sufficient knowledge and appreciation of the modern approach to quality and reliability to give the right lead, and to take the right decisions with regard to the company's quality policy which are not always easy to base on clear financial estimates. Middle and lower ranks of management are quick to spot whether interest and appreciation at the top is genuine, and there is no question that those companies which have had the greatest success in enhancing their reputation and expanding their sales on the basis of reliability and value for money are those where the chief executive has taken an effective personal lead.

In the national campaign for quality and reliability in 1966–7, great stress has been placed, quite rightly, on the phrases *total quality control* and *quality is everybody's business*, and this has helped to modify the view, possibly encouraged by much recent literature, that the mere setting up of specialized quality control, quality assurance or value engineering departments will quickly achieve the desired ends. Nevertheless, it is important to avoid the danger that in quality becoming everyone's business it does not in practice become nobody's business, and to recognize that in this technical age we must have specialists and specialist departments, without which and without whose guidance we are bound to miss using many of the best methods and techniques, and fail to realize how far we are falling below what we might achieve.

A very important point in organization for improved quality control is to recognize from the outset that we are providing a service to help existing staff to shoulder and carry out their responsibilities better and not introducing a new man or new department to take them over. Cowan, in his book *Quality Control for the Manager*,[9] points out a danger in the very name quality *control*. Control implies management and responsibility, and one often hears the question, 'How can I be held responsible for something I do not control?' The name, however, is so well established both nationally and internationally that it may be difficult to avoid its use. If we are making an appointment with the object of developing and improving this aspect of our business, Manager, Quality and Reliability Services, might be a better title than Manager, Quality Control. Titles, however, are of secondary importance to acceptance, and what we are primarily concerned with in any new appointment is preparing the way for success in achieving our objectives. Quality control is not an end in itself, but a means to an end, and that end will be best achieved if line management fully accepts its responsibility for turning out goods of the required standard, and welcomes and values the help that specialists can give towards achieving these aims.

Before we go on to discuss the organization of a comprehensive quality and reliability set-up and the various specialists who may be required, let us consider the case of the small or medium sized firm or division where it has been decided to improve and modernize quality and reliability services. There will be an existing set-up for control of quality, probably an inspection department headed by its chief, and designers, line management and supervision all held responsible to varying degrees for the quality of their work. There is, however, no specialist service responsible for co-ordinating efforts and techniques directed at obtaining the best results in the most economical way. Records and data on quality aspects may be unsystematic, and while the firm may be turning out a lot of good work, rejections and scrap may be high. It has been decided to make an appointment with a view to improvement. How do we choose the individual and how should he tackle the initial stages of his job?

A knowledge of the firm's problems and techniques is a great advantage, and it will be best to look first inside the existing organization. It will, however, almost certainly be impossible to find someone who is experienced in all processes to which he will, in due course, be providing a service. First and foremost we must choose someone with the right personal characteristics, an ability to collect facts without prejudice and to communicate them to others, and build up a sense of partnership and co-operation.

A second qualification which will be of great value is a sound technical training in engineering or science. An ability to look at matters with a scientific attitude and to be able to express facts in the form of figures will be immensely valuable. Thirdly, some factory experience and ability to understand and get on with factory personnel will be important. Training in statistics, while useful, is less essential as an initial qualification. Such training will almost certainly be necessary, and can be usually acquired fairly quickly taking advantage of education facilities and courses which are now widely available. As the new service develops, and the part which statistical methods will play becomes clear, the appointment of a specialist in this line may become desirable.

Having made the appointment, the first jobs to be tackled are probably not the most serious quality problems, which readily spring to light. The best start will be made by choosing an area where quality data can be easily classified and measured without too much disturbance to existing methods. In the author's experience in a floorcovering factory, the first job tackled after appointment of a quality control specialist was thickness of linoleum tiles. This was by no means the company's most serious quality problem, but the success obtained and savings made in this investigation both gave confidence to the newly appointed officer and convinced the manager and foremen concerned of the value they could obtain from the new service.

Now that the new organization has started and begun to obtain acceptance from management, work can be expanded and new problems tackled. At this stage we can begin to handle more major problems. Contrary to what is often imagined, it is usually best not to start at the beginning of the process, raw material supplies, and work systematically through the working procedures, but to start with end effects, find out where troubles, complaints and high rejections lie, and then work back from effects to causes. In this way valuable effort can be saved which might otherwise be dissipated on bringing about changes which have little effect on the final product and cost of manufacture. In the course of such investigations it may well be found that raw materials and incoming parts are an important factor in the troubles we wish to avoid, and the time is then ripe for tackling acceptance procedures and developing better liaison with suppliers. Similarly it is far better to take a problem and, after analysing it, consider the techniques most likely to give a solution or an improvement, than to start with an interesting technique and look for possible applications.

We now come to the stage at which the improved organization has begun to justify itself by results and to be accepted and appreciated by management and supervision. Broadening of activities may now be required, so that advantage can be taken of improved techniques over a wider field. The time has come for the appointment of additional specialists and for training and education in the company as a whole. Let us therefore give some attention to the question of quality and reliability specialists, the part they can play, how they can best fit into the organization structure and how training should be developed.

Statistical quality control
Two chapters in this book have been devoted to statistical methods and their importance, particularly in quality of conformance. Statistical quality control should play an important part in all industries where quality conformance is capable of measurement, and they comprise the great majority. Statistical methods must be widely applied, understood and trusted if they are to have their maximum effect. Professor Grant, in his book *Statistical Quality Control*, stresses that there are four levels of training and knowledge required in an organization effectively using these techniques.

(1) The level of understanding the mathematics on which statistical methods are based, and ability to read and understand the literature of mathematical statistics.
(2) General understanding of the principles underlying

control charts and sampling tables, why they work, how to interpret results and to decide on the method to use in any particular case.

(3) Broad understanding of objectives and possible uses of statistical quality control, without necessarily sufficient detailed understanding for close supervision of the work. This is particularly valuable at higher management levels.

(4) Ability to use one or more of the techniques on a rule of thumb basis. This applies particularly to routine inspectors and machine operators.

Wherever there is scope for comprehensive use of statistical methods, we require at least one specialist at the first level. Without this we are bound to fall behind with modern developments and fail to use to full advantage the best available techniques. The status and responsibilities of such a statistician will depend on circumstances. He may have executive authority, for instance as chief inspector, or a senior position in the laboratory. Whatever his position, he needs not only technical ability but good personal relationships and acceptance by those who need his specialist advice and guidance.

Given the availability of the specialist, the success of application will depend very largely in bringing as high a number as possible into class (2), and having them widely distributed among departments. The more departmental managers and supervisors on production and inspection, methods and design engineers and control chemists that can be brought into this group, the more successful will the programme be. This requires a training programme in which the specialist or specialists should play a part, but which can probably be assisted by courses at the local technical college or elsewhere. In industries such as plastics, papermaking, rubber and textiles, much of the testing control and process improvement is laboratory work, and a wide knowledge of statistics at least on the second level is desirable among laboratory staff.

One type of organization which may be extremely effective is to have a small statistical quality control department, headed by a suitable specialist, with perhaps two or three assistants who will help and advise departments and carry out studies wherever required. Such a department can often give a service which makes possible returns and savings out of all proportion to its cost.

Value analysis or value engineering

It is generally recommended in the literature and by experienced consultants that for the successful organization of value analysis or value engineering there should be a full-time specialist as chairman or leader of the team. The other members of the working group should remain in their departments and carry their normal responsibilities. Companies intending to start value analysis for the first time may be well advised to employ consultants to introduce the technique, conduct the first exercise and train the proposed value analyst. The alternative is to send a selected member of the staff, designated for the position of value analyst, on a suitable course, at which the delegate may be able to take a product or process from his own company and have it submitted to value analysis under expert guidance.

Inspection

It has been stressed throughout this book that inspection is not synonymous with quality control and that the view, sometimes held, that responsibility for ensuring product quality should rest fairly and squarely on the shoulders of a chief inspector and his staff is likely to be costly and inefficient. We have seen the importance of line management and supervision fully appreciating and accepting its responsibility for the quality of the goods produced, and of avoiding the conflict of interest which so frequently occurs between producers and inspectors, particularly at shop floor level. The dogma that a man cannot be trusted to inspect his own production, that a foreman should never be responsible for supervising the inspection of the goods his men have produced, or even that a departmental, production or works manager is likely to be prejudiced and should not be responsible for his inspection, is still quite widely accepted and practised. The foreman who has been given responsibility for inspecting production and who deliberately lets through faulty work for which he is also responsible will, in any well-managed factory, do it only once. The day of reckoning when the material is rejected by the next process or by the customer is inescapable and the lessons not easily forgotten.

Inspection, even if the term is often too widely used, is an important and usually essential function and many of the best and most efficient methods today are so technical that it is essential for a suitably qualified specialist to be in charge. This applies equally in the engineering industry, where for instance non-destructive-testing methods such as radiography and ultrasonics require expert technical administration, as in the chemically based industries where many of the analyses and tests may require laboratory control and conditions. This does not necessarily imply that all such inspection functions need be in direct line authority to the expert, but it is important that he should be responsible for ensuring that the inspectors are efficiently trained, that results are correctly interpreted and used, and that the system is technically maintained. In many situations it is a good plan for all straightforward on-the-line testing, such as gauging and measuring, to be the direct responsibility of the production foreman, and for the inspection department as such to be limited to a comparatively small number of technical and sophisticated methods.

It is obviously impossible to lay down any general rules as to how the chief inspector should fit into the organization plan and to whom he should be responsible, for example whether he should report to a technical director, works manager, or to a head of management services. This must depend on the nature of the industry, the qualifications and abilities of the individuals concerned and on general company organization policy. What is important is to cultivate a relationship whereby the chief inspector and his staff provide a service which is valued and appreciated by production management and supervision, even if he may not be responsible to the production chief.

It will also depend on circumstances as to whether the duties we discussed above of head of statistical quality control and chief inspector can be combined in a single individual. Where circumstances permit and a man can be found or trained to have the necessary technical and personal abilities, this may be a very good arrangement, especially in a small or medium sized organization.

Apart from the position and responsibilities of the chief inspector, a very important feature is the selection and training of the inspectors themselves, their status and method of payment. Too often the selection is very arbitrary and it is not uncommon for a man who has, on account of age or infirmity, proved to be too slow for production to be drafted into inspection. Inspectors are often worse paid than the men whose work they inspect, because of the effect of production bonus, which is paid to the producer but considered wrong or dangerous to pay to an inspector. This is rather an illogical attitude; the producer has as great a responsibility for making the article right as the inspector has for seeing that it has been made right, and if production bonus is likely to cause neglect or carelessness in one function the same will almost certainly apply to the other. It is far better for the on-the-line inspector to be a full member of the production team with similar pay conditions and trade union status, so that no difficulty occurs when transfer or interchangeability is required. If the idea of paying production bonus to a process inspector is regarded as too heretical, the only reasonable alternatives are to replace production bonus by high day rate or measured day work, or to give the inspector staff status, or other advantages which will make his job no less acceptable or desirable than that of the production operative. Once this obstacle has been removed, the problem of selecting and training the right man or woman for the job is greatly simplified and the whole standard of reliability can be improved. Uniform conditions of payment and status can be of special help in visual inspection. Inspection fatigue, as we have seen, has a very quick effect on the efficiency of a visual inspector, and it is very desirable where possible to give frequent changes of job to those concerned. In many processes, provided pay and status conditions are uniform such interchange between producers, inspectors and packers is easily arranged. Where this is impossible due to production layout, arrangements should be made to give variety in the nature of inspection during the shift.

In some industries where quality and reliability not only form a very important part of company policy, but where the technical considerations involved are considerable, it is desirable to have an autonomous department responsible for the measurement and control of quality and reliability, headed by a manager of high standing in the company structure, who may be an executive director, or at least responsible directly to the board. This department may cover quality engineering, testing of raw materials and control of bought out components, machine capability studies, inspection, quality investigation and reliability testing. Service reports and complaints are all channelled back through this department, co-ordinated and fed back to the departments concerned, so that the product can be improved both in design and conformance to customer requirements. This type of organization has proved very successful in many companies with a high reputation and image for quality and reliability. The essential feature is that the design, production and service departments also fully accept their own responsibility for conformance with the standards laid down. If a production department possesses the attitude that its responsibility is limited to turning out the scheduled production and that quality is purely the concern of the quality control department, the whole essence of the philosophy fundamental to the modern view of quality and reliability control breaks down.

In conclusion, it will be useful to consider one or two further aspects of organization which may help towards the success of a programme for improved quality and reliability.

Location
This is a much more important factor than is often realized. The success of a quality control department greatly depends on acceptance and co-operation from the shop floor, which is, after all, where quality is made. The more production personnel see of the quality control staff, the less suspicion there will be and the more readily will they be accepted as partners. Wherever possible, therefore, the staff should carry out their work directly adjoining a production department, or even, as described in a case study in Chapter 2, alongside the actual machine. Another advantage of close location to production plant is the speed of feed-back of information, which is so vital in our objective of making material right first time rather than classifying and sorting it later on.

Prospects for quality control staff
The attraction and retention of good staff is of course a major factor in success, and the prospects of progress and promotion play a major part in this. Quality work is an excellent training ground for other positions in the business. It brings a man into contact not only with production and design, but with suppliers and customers, teaches him how to get on with and get results from other people, and develops the systematic and logical approach so useful to the senior manager. The success of a quality control department will be aided if company policy uses it with this aspect in view. The temporary inconvenience resulting from promoting a useful member of the quality control team to a management position will be quickly repaid by the calibre of the new staff it becomes possible to attract.

Quality and budgetary control
In most modern manufacturing plants, pressure for the reduction of costs and prevention of cost increases is exerted by budgets, and it is a principle of budgetary control that such pressure, to be effective, must be applied to individuals and follow the lines of individual responsibility. Costs related to quality are such a fruitful field for savings, and so frequently cut across the lines of departmental responsibility, that it is important to ensure that our organization for quality and reliability and our budgetary control system are co-ordinated in such a way that there is no obstacle to progress.

Often a small increase of cost in one department may make possible a much larger decrease in another, or an apparent economy may be at the expense of seriously increased costs elsewhere. Improvements in the manufacture of components may increase their production cost, but give a much greater saving in assembly. A re-organization of goods acceptance procedure may have an important effect on production costs. Where production and inspection are under separate management, the introduction of statistical methods in process control, such as charts for measured variables, may, at all events temporarily, increase production costs while resulting in a greater saving in the cost of final inspection. What matters in cases of this kind is the maximum benefit to the organization as a whole.

The authority for quality decisions, therefore, should normally rest at a management level higher than the supervision directly involved. The foreman with his intimate knowledge of the job and his direct authority for the operative can, none the less, play a most important part in promoting improvements, and it should be an important feature of the training and communications system to put across to supervisory staff the broader view of total savings.

Credit for quality improvements
The question as to who gets the credit for quality improvements and cost savings should be an unnecessary one, but human nature being what it is, is apt to assume considerable importance. Cost savings and an improved product are obtained by team work in which designers, line management, supervision, machine operators and the quality technician all play their part. Nothing could be more fatal to success of the organization than an attempt to allocate credit exclusively to one partner. When the adoption of a changed procedure results in an improvement, it is important that both the originator of the idea and the man who makes the change work in practice feel that their contribution has been recognized.

To sum up, the type of organization for quality and reliability which will ensure a sound reputation for giving the customer good value for the money he spends depends on the nature of the business and the personalities of those concerned. What is vital to success is first a clear policy laid down and actively supported by top management, secondly adequate technical staff and organization to ensure the best available methods of control are used, thirdly trust and confidence between those who control and those who design and produce, and finally, and perhaps most important, an acceptance of their own responsibility in carrying out the policy by both specialists and non-specialists alike.

8 Economic Aspects of Quality and Reliability

'Prior to World War II, Japanese products were notorious for shoddiness. For post-war Japan, to whom exports were its very lifeblood, this situation was intolerable. The Government therefore helped to establish export standards which now compare with any in the world.'

Shinzo Ohya
Quoted in *The Times Business News*
July 17th, 1967

THROUGHOUT THIS BOOK it has been stressed that a sound policy and organization for quality and reliability is not only good ethics but good business economics. Money is the only variable which can be applied to measure in comparable terms a company's varied activities, and the most generally accepted criterion for management decisions in an organization working for profit is profit maximization. When we attempt to deal with decisions affecting quality on the basis of economics, we are sometimes faced with a more difficult problem than in such matters as investment of industrial assets or studies in production methods because of the difficulty of expressing in financial terms the effects of quality decisions. This difficulty leads to a tendency to decide quality matters on the basis of intuition, or worse still to evade decisions and fail to take action because the effect on cost and profit of a quality proposal is not clearly stated. We have seen that a major factor in improved quality control is an ability to express quality in terms of measurement. Where we can extend this principle to measurement in the universal standard, money, the decision-making process, particularly as it affects other activities of the company, is greatly assisted.

This can be simply illustrated by a reference to tolerances. One of the important techniques of value analysis, listed in Miles's book,[10] is *get a dollar sign on tolerances*. Let us suppose that a component which is required for an assembly has been based on the qualitative standard 'it must be a good fit'. The first step forward in quality control is made when the dimensions which will give a good fit are expressed and measured in terms of a specification mean and tolerances. Instructions to the maker and inspector can now be precisely stated and we know much more clearly what is required. When we come to the second step and measure in money terms what working to these specified tolerances is costing, further progress can be made. We can now decide whether we are getting good value for our efforts and whether modified tolerances or some alternative method of achieving our aim more economically is possible.

We shall, therefore, discuss to what extent quality costs can be extracted and classified and can be used to help logical decisions on a number of quality matters. At the start of Quality and Reliability Year in 1966, stress was placed on the high cost to British industry of scrap, rework and faulty material, which was estimated to be costing the country at least £650 million per year. Quality costs were defined as the combined costs of prevention, appraisal and failure, and it was estimated that in a wide range of industry these costs often represented 10% or more of a company's turnover. It was urged that if we spent more on prevention much greater savings would result in appraisal and failure costs, and Figure 8.1 illustrates this principle diagrammatically.

Cost of prevention includes the salaries and expenses of staff specifically engaged in quality engineering and quality control, the cost of research, investigation and trials on quality improvement, and the cost of operative and staff training in quality aspects, but excludes the cost of routine inspection of incoming or finished goods, which are classified as appraisal costs.

Failure costs normally represent by far the largest proportion of quality costs, and may be conveniently subdivided into production costs covering the cost of scrap, rework and discounts

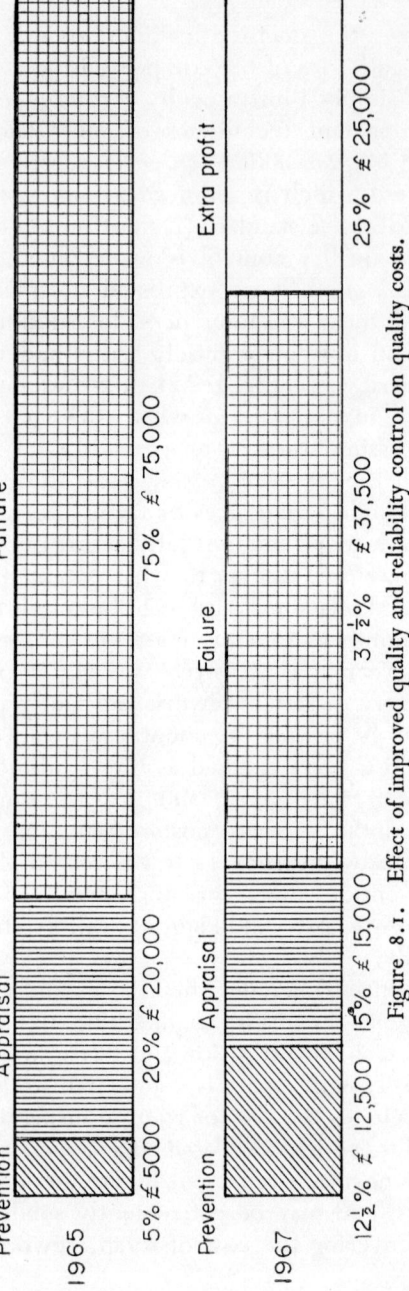

Figure 8.1. Effect of improved quality and reliability control on quality costs.

on jobs or seconds resulting from manufacturing faults, and unsatisfactory product costs, which include the cost of investigating and settling complaints and any other expenses or loss of revenue arising from quality defects in the product. Use of costs of this type help in the making of logical decisions on such matters as the following:

(1) Should we employ 100% inspection, sampling inspection or no inspection for incoming goods, intermediates and finished products? If the choice is in favour of sampling inspection, what is the most economical AQL?
(2) Should patrol inspection be introduced during manufacture, with a view to reducing final inspection and failure costs?
(3) Should process quality control or finished goods inspection be on the basis of measured variables instead of attributes and will it pay to use control charts for variables? Here there is often a great opportunity for cost saving, since while the cost per test will be more where measurement is involved than by simple go-no-go methods, the number of tests may be substantially reduced and the extra information obtained and fed back may lead to great improvement in scrap and rework.
(4) Is the inspection of a component justified, or is it more economical to allow the fault to pass forward and be rectified after testing the final assembly?

Apart from aiding decisions of this nature, quality costs are of value in verifying whether introduction of new methods of control have been justified by results. In particular, the question may be asked a few months after statistical quality control methods have been newly introduced as to whether management is satisfied that their continuation is justified. Sometimes the position may be quite clear as a result of a marked drop in spoilage or rework or economy in the number of inspectors employed, but frequently during a change of this kind so many other factors are involved that a comparison of results before and after the event may require great care in interpretation.

Another important aspect of quality costing applies to design improvement and value analysis. As in our example on tolerances given above, the whole basis of value decisions depends on obtaining realistic costs and cost comparisons between alternatives. Many of the savings which result from such investigations lie in the field not normally regarded as that of quality costs, the use of cheaper materials, reduction in labour costs through simpler assemblies or more realistic tolerances, and the economies resulting from purchasing outside rather than making in or *vice versa*. None the less, such costs affect important quality decisions, and by treating such investigations individually it is easy to compare the cost of conducting the study against the total cost savings which are obtained.

Thus we see that the extraction and classification of quality costs can be a great help in decision making and a stimulus for improvement. It must not, however, be forgotten that the greatest effect which improved attention to quality and reliability may have on a company's profitability will probably be the long-term one of expanding business as a result of customer satisfaction and improved reputation. This aspect is less easy to estimate and express in clear financial terms, and furthermore is usually concerned with expenditure which must be written off in the current year against the prospects of growth in later years. The position is comparable with the justification of expenditure on promotion and advertising with a view to future sales. Like advertising, good quality control is a thing we cannot afford to do without.

It may be relatively easy to assess the cost of settling customer complaints, or replacing goods which fail during a period of guarantee, but putting a money value on the effect on the company's image and reputation and in turn on future sales, progress and profit is a much more difficult problem. The short-term view of profit maximization if applied rigidly to all decisions affecting quality and reliability may well conflict with the longer-term interests of both the company directly and the national economy, on which individual companies in the long run must depend. The attitude that it is unnecessary to devote money and time to improvement of quality and reliability because good profits can be made without such efforts which was encouraged by the seller's market after the last war is still far too common in industry, and may in the long run prove disastrous. Neglect of planning for, and investing in, quality and reliability will have the same result as neglecting research and development, or lack of planning for the future in the engagement and training of staff; short-term profits may be satisfactory, but there will be the inevitable day of reckoning in the future. Fortunately, as we have seen, if we go about our quality control in the right way direct savings, particularly in the cost of scrap and rework, often more than pay for the extra costs involved even in the short term. Where, however, it is difficult to prove in figures that savings will exceed additional expenditure, it may still be short-sighted and unsound to reject a policy which will lead to a reputation for good quality, reliable products and services and value for money to the customer.

Before discussing this problem from the viewpoint of the individual company, let us look at it from the national and international aspect. In Great Britain, following the immediate postwar period with its shortage of goods and seller's market, the economy has been subject to recurring cycles of about five years. A period of high home demand and consumption is followed by one of difficulty with balance of payments and the need for retrenchment. Trade slumps and high unemployment such as we experienced before the war have been avoided, mainly due to the rapid advance of technology and the partial recognition of the economic fact that a prosperous economy demands a prosperous market. Nevertheless, these five-year cycles continue to occur and national progress in productivity has lagged behind many other countries. The quality and reliability factor has been by no means a minor item in this situation. During boom periods home demand is high, imports increase, and wage spirals and shortage of skilled workers result. The natural consequence, particularly where industries are concentrating on immediate profits, is neglect of quality control, late deliveries and unreliable service, all of which combine to bring about the next period of difficulty. If National Quality and Reliability Year, which coincided with the recent balance of payments crisis, has a progressive and lasting effect on the attitude of British industry it will have played an important part in avoiding, or at all events minimizing, the severity of such crises in the future.

The best example in the international field where a modern attitude to quality and reliability is paying dividends is probably Japan. A high proportion of Japanese industry has adopted enthusiastically modern quality and reliability policy and the greater part of their exports are subject to sound quality control. As a result, a country which before the war had a reputation for cheap goods and imitations has risen to the forefront as a supplier of value for money and reliable goods and services. This has played a major part in Japan's remarkable expansion in productivity and the growth of her international trade.

We shall now look at the matter more from the point of view of the individual company. What are the company's objectives? Statements of objectives vary from firm to firm, and though profit maximization is the one most frequently cited, in actual fact it is rarely the sole aim. Other subsidiary objectives may include such matters as growth, reputation, the well-being of employees and service to customers. F. W. Taylor defined the objective of good management as the maximum prosperity of the employer coupled with the maximum prosperity of each employee in the organization. Whatever the individual objectives, the best summing up, which few would dispute and which applies to most, if not all, of the best concerns, is that we want to do a good job and do it efficiently. To do the job efficiently we must be profitable, and profitability is therefore always an essential objective of any well-managed company.

At a recent conference organized by the Production Engineering Research Association, Professor Ezra Solomon of the Graduate Business School, Stanford University, stressed the difference between what he described as profitability, the ability of an undertaking to produce or create new wealth, and profit maximization as it is generally understood. The following quotation is taken from his paper.[11]

> 'Rather than profits, a more useful operational concept is what may be called profitability. This is defined as the ability of an undertaking or of an economic unit to produce or create new wealth. The selection of the wealth maximization as a central objective does not imply that the firm has no other motives or objectives. Clearly, management needs to

ECONOMIC ASPECTS OF QUALITY AND RELIABILITY 117

be concerned with a broader range of considerations beyond the scope of finance, as well as with sub-objectives such as sales, size and growth of market just to name a few. But as far as resource-using decisions are concerned, wealth maximization provides a meaningful operating objective, which is generally consistent with and indeed complementary to the attainment of other goals.'

If we look for a company whose progress and success can be measured by growth and the production of new wealth, an outstanding example is Marks and Spencer. Between 1956 and 1966, turnover increased from £119,000,000 to £238,000,000, and net profit before tax from £10,099,000 to £29,818,000, advances in each figure being shown in every successive year. The company's attitude to quality and reliability and the influence it has exerted on manufacturing suppliers to provide dependable products has undoubtedly played an essential part in this successful record. The following extracts are taken from a paper given at a conference on 'Profiting by Quality and Reliability' in Blackpool in November 1966.[12]

> 'The quality policy of Marks and Spencer was developed and laid down by the late Lord Marks and is being continued by our present chairman, Lord Sieff, and the board. In brief, the policy is to give the customer real value for money. The trade mark of "St. Michael" has, over the years, become synonymous with quality merchandise in both food and garments. This is the result of the fundamental philosophy of the company that quality is the basis of our business and is achieved only by constant vigilance. Our customers expect not less than the best value for their money.
> 'Our policy of constantly upgrading our merchandise has imposed on our suppliers unceasing demands to meet these very exacting specifications.
> 'We do not own or control manufacturing concerns. In order to achieve consistent and interchangeable standards, specifications for all of our merchandise have been developed which lay down in detail all relevant factors, e.g. sewing threads, seaming details, colour fastness, etc., in textiles and

E

recipes detailing quality and purity, etc., in foods. These specifications are the heart of our business. Our suppliers, over the years, have fallen in with us and are as conscious of the necessity of producing only quality merchandise. This applies equally to food and textile products.

'First of all, it should be pointed out that Marks and Spencer is basically a team and that, from the board downwards, we all work together towards a common goal. Therefore the idea of quality, and the determination to achieve quality, permeates every aspect of our business. Our directors continuously scrutinize merchandise and actually wear, and get their families to wear, our garments and eat our food from our counters. If there is the slightest doubt as to the quality, fit, appearance and, in the case of food, taste and palatability, these complaints are communicated to the executive in charge of the particular group who, of course, then takes immediate action.

'Manufacturers come to us with manufacturing problems of almost every kind and we then try as a team to help them overcome these difficulties. But not only the technical staff are concerned with quality. Selectors (buyers) and merchandisers (distributors) continuously and frequently visit suppliers' factories to ascertain for themselves that the garments or other merchandise adhere to the specifications. Should any deviation or deterioration be found, these will be eliminated in discussion with both top and factory management of the supplier. Should technical difficulties be involved, the appropriate department will be asked for assistance and advice.

'We advise our suppliers on layout, flow of production and on the use of existing machinery. We help them in developing new methods and equipment and keep them continuously posted on new devices both from the United Kingdom and abroad.

'All the above has not been achieved suddenly (it was started after the war by the late Lord Marks) and a relationship has been built up between suppliers and ourselves which is based on mutual trust and an understanding of each other's difficulties and aims.'

ECONOMIC ASPECTS OF QUALITY AND RELIABILITY

Many examples could be quoted of manufacturing companies both at home and abroad who have good quality and reliability policy and have shown a steady growth of sales and profits over a period of years. The following figures for two companies well known as exponents of modern quality control serve as an illustration.

	Sales × £1000		Net profit before tax × £1000	
	1956	1966	1956	1966
Metal Box Co., Ltd.	61,080	141,442	5608	13,111
	1957	1966	1957	1966
Joseph Lucas Limited	84,500	187,200	6782	11,778

Of course many other factors have played a part in the obtaining of these successful results, but it will be generally agreed that without a sound policy of quality and reliability and the creation of the resulting image and reputation for good value for money such progress could never have been realized.

While many firms have good systems for control of quality and reliability, unfortunately some manufacturers fail to measure up to these standards, as can be illustrated by the following quotations from a paper by the Chief Inspector Engineer of the Crown Agents, London, referring to cars for export.[1]

'Minor faults are found, such as paint oversprayed, paint chipped, chromium plating scratched, blowing joints, windows stiff, etc. Chassis frames are found covered in mill scale and spring brackets have been seen to be cracked. Engines are usually reasonably satisfactory although minor oil leaks have to be corrected and firms sometimes forget to fit the correct tropical fan. Gear boxes give more trouble than would be expected and, not infrequently, have to be changed for one fault or another. The same applies to rear axles, where noise is a frequent cause of rejection. Brakes are, quite often, found on test to be incorrectly adjusted and needing to be put right, while electric accessories give a variety of trouble.

... It is to be regretted that suppliers of vehicles take a tolerant view of their own shortcomings. One prominent firm has even gone so far as to say that it is not in business to supply finished vehicles off the assembly line, but only "completed assemblies of parts". . . . There is thus a danger that the manufacturer will attempt to pass to the customer's inspector his responsibility for supplying correct quality and this has to be resolutely guarded against.

'At one of the largest vehicle manufacturers in the country, where the inspector has firmly declined to accept unsatisfactory vehicles, it has now been arranged that vehicles for the Crown Agents are specially labelled on the assembly line so as to secure preferential treatment; a striking admission of the value of independent inspection, but a sad confession of the failure of quality control of the standard product.'

A factor which is sometimes not appreciated is the relationship between quality and reliability and productivity. If we follow the principle that the efficient method of obtaining a reliable product is correct design for the purpose intended and effective steps to see that manufacturing processes can and do conform to the requirements of the design, productivity goes hand in hand with the enhancement of quality. The elimination of factors in design which contribute nothing to the function and attractiveness of the finished products, studies to ensure that our plant is capable of producing the product required within the permissible range of tolerance, avoidance of the delays involved in correcting errors, and finally reduction in the time spent sorting good from bad in final inspection, all result in more useful output from each man we employ. Great stress has been placed on the importance for the national economy of a steady growth rate in the productivity index. What is sometimes overlooked is that increased productivity requires two things: increased production of saleable goods and increased demand abroad and at home to absorb the increase. The development of a modern attitude and approach to quality and reliability over a much wider area of industrial manufacture and services could do more in achieving these aims than any other single factor. Work study, productivity agreements, mechanization and automation can all help to produce more; good promotion and advertising and salesmanship can help to sell more, but the right attitude to the problem of producing reliable goods without unnecessary costs and giving the customer real value for the money he spends serves both ends. Without this approach all other efforts, essential as they may be, are doomed to failure, or at the best, limited success.

Books Recommended for Further Reading

M. J. Moroney, *Facts from Figures* (3rd Edn.), Penguin, Harmondsworth (1956).
A good introduction to the basic methods of statistics, with many practical examples. Demands only a knowledge of elementary mathematics.

E. L. Grant, *Statistical Quality Control* (3rd Edn.), McGraw-Hill, New York (1964).
A comprehensive working manual on the detailed techniques of statistical quality control, covering practical and theoretical aspects.

A. V. Fiegenbaum, *Total Quality Control*, McGraw-Hill, New York (1965).
Another first-class book, covering the wider aspects of quality control, with special attention to quality costs. Examples are mainly from the engineering industry.

L. F. Thomas, *The Control of Quality*, Thames and Hudson, London (1965).
A book essentially for the practitioner, shorter and less comprehensive than the McGraw-Hill books listed above. The basic statistics are clearly explained and special consideration given to human aspects.

A. F. Cowan, *Quality Control for the Manager*, Pergamon, Oxford (1964).
A short, easily readable introduction to the philosophy and practice of quality control.

O. L. Davies, *Statistical Methods in Research and Production* (3rd Edn.), Oliver & Boyd (for I.C.I. Ltd.), Edinburgh (1957).

A well-written textbook outlining the basic methods of statistical analysis including sampling and quality control, with many practical examples.

R. H. Woodward and P. L. Goldsmith, *Cumulative Sum Techniques*, Oliver & Boyd (for I.C.I. Ltd.), Edinburgh (1964).
A useful monograph on how to use cu-sum charts for quality control and other applications.

Lawrence D. Miles, *Techniques of Value Analysis and Engineering*, McGraw-Hill, New York (1961).
The standard textbook on value analysis. Excellent reading, with a clear account of the techniques and numerous practical examples.

C. Hearn Buck, *Problem of Product Design and Development*, Pergamon, Oxford (1963).
A good clear introduction to engineering design and the principles involved in giving reliability at an economic cost.

J. M. Juran, *Quality Control Handbook* (2nd Edn.), McGraw-Hill, New York (1962).
A comprehensive and useful reference book. Probably the most complete source of information on quality control available today.

Periodicals

Quality. Journal of the European Organization for Quality Control (via the British Productivity Council or National Council for Quality and Reliability).

Industrial Quality Control. Journal of the American Society for Quality Control.

References

1. W. D. Farrington, 'Overseas Contracts', Conference, Blackpool: Profiting by Quality and Reliability (1966).
2. F. W. Taylor, *Scientific Management*, Harper, New York (1947).
3. E. L. Grant, *Statistical Quality Control*, McGraw-Hill, New York (1964).
4. G. R. Gedye, *Scientific Method in Production Management*, Oxford University Press (1965).
5. Her Majesty's Stationery Office May 1961, *Sampling Procedures and Tables for Inspection by Attributes*.
6. M. T. Witts, *The Production Engineer*, April 1967, Figure 4, page 247.
7. M. T. Witts, *The Production Engineer*, April 1967, Figure 6, page 249.
8. M. T. Witts, 'Better Engineering with Simple Statistics', *The Chartered Mechanical Engineer*, October 1966.
9. A. F. Cowan, *Quality Control for the Manager*, Pergamon, Oxford (1962).
10. Lawrence D. Miles, *Techniques of Value Analysis and Engineering*, McGraw-Hill, New York (1961).
11. Ezra Solomon, *Financial Planning*, PERA Conference, Melton Mowbray, May 1967.
12. H. L. Fiebelman, 'Retail Distribution', Q.R.Y. Conference, Blackpool: Profiting by Quality and Reliability (1966).

Index

ACTION LIMITS 59
Attributes
  control by 25, 61–64
  inspection by 25, 39
  sampling by 38

BUDGETARY CONTROL 108

CAPABILITY STUDIES, PROCESS 32, 33, 83
Car industry 78, 119
Company objectives 116
Company policy in quality and reliability 106
Competitions 20
Conflict of Interest 8, 16–18
Conformance, quality of 32, 76
Control charts 46–74
  attributes 25, 61–65
  cumulative sum 68–74
  measured variables 25, 44–61
  mean and range 51–57
  percent defective or percent yield 65–68
Costs
  appraisal 111
  failure 111
  prevention 111
  product 77
  quality 1, 111–115
Critical defects 14, 15
Crown Agents 119 120
Customer, quality control and the 2, 76, 114, 116
Cumulative sum charts 68–70

DEFECTS
  classification of 14, 78
  critical 14, 15
Design 75–97
  improvement 113
  quality of 2, 75
  time factor in 90

EXPORT TRADE 78, 116, 119–120

FAULT CLASSIFICATION 14, 78
Feed back of information 10, 77, 79–80
Finished product performance 20
Frequency distributions 29–37
Function 88

GENERAL ELECTRIC CO. (USA) 97
Goods inward inspection 38, 40, 41
Government contracts 76
Grant, E. L. 39, 40, 102

HAND-CONTROLLED OPERATIONS 55
Histograms 31
Hunting 50

IMPROVEMENTS, RECOGNITION AND REWARD OF 19, 23
Incentive schemes, their relation to quality 16, 18
Inspection 3, 8, 104–6, 111
  and quality control 6
  100% 11, 113
  sampling 11, 113
  visual 12, 13, 106

Inspectors
  conflict of interest with producers 8, 16, 18
  selection and training of 106
Integration of inspection and production 14

JAPAN, ATTITUDE TO QUALITY AND RELIABILITY 116

LIMITS AND TOLERANCES 46, 47, 85, 89, 111
Location of quality control department 107
Lower control limit 65
Lucas, Joseph 97, 119

MACHINE-CONTROLLED PRODUCTION 55
Management
  line 104
  responsibilities 17, 100
  top 5, 109
Material specification 88
Marks & Spencer Ltd. 117
Mean 26
Measured variables 25, 39, 41, 113
Measurement of quality 25, 111
Metal Box Co. Ltd. 119
Method study 88
Miles, L. D. 85, 111

NATIONAL ECONOMY, EFFECT OF QUALITY AND RELIABILITY ON THE 114–115
National Quality and Reliability Year 1, 115
Normal distribution 27, 28, 29, 47

P–CHART 65–68
Paper making industry 64, 70–72
Patrol inspection 113
Piece-rate 7
Planning for quality and reliability 114, 115
Plastics Industry 10, 11, 64

Poisson distribution 62
Premature failure 78
Pride of workmanship 18, 20, 23
Process capability studies 83
Product cost 77
Production engineering 83
Production, integration with inspection 14
Productivity, relation to quality and reliability 120
Profitability 1, 116

QUALITY
  bonus 17
  control staff, prospects for 108
  control staff, selection 100–102
  costs 1, 111–114
  incentives 7, 17, 18
  of conformance 2, 24, 102
  of design 2, 24

R–CHARTS 51
Random samples 38
Reliability 2, 81, 83
  testing for 79, 83
Rover Co. 78, 79

SAMPLE AVERAGES 53
Sampling inspection 11, 113
Sampling theory 27, 37, 41, 44
Semi-automatic control 59
Shewhart, W. A. 27, 74,
Shop floor presentation 20
Significance 27
Significance tests 28–29
Skew distributions 30
Small firms 100
Soloman, Ezra 116
Specifications 3, 46, 75
Standard deviation 26, 66
Standardization 88
Statistics 4
Statistical methods in quality and reliability 24, 45, 102
Stratified samples 40

INDEX

Staff selection for quality and reliability 100–102
Supervision 104

TALLY CHARTS 29
Taylor, F. W. 16, 116
Textile industry 36, 64
Time factor in design 90
Tolerances 47, 85, 89, 111
Total quality control 5, 99
Top management 5, 109
Training 17, 18, 19, 102, 103, 115

VALUE ANALYSIS 85–97, 104, 114
Value engineering 85, 104
Variability 3, 4, 24, 26, 32, 37, 42
 inherent 48, 66
Vendor rating 41
Visual inspection 12, 13, 106

WASTE OF MATERIAL 65, 88
Warning limits 59, 66
Weaknesses, inherent 78

$\bar{X}$–CHARTS 51